T0296843

Alzheimer's Disease Theranostics

Alzheimer's Disease Theranostics

MAGISETTY OBULESU
Scientist, ATG Laboratories, Pune, India, Certified by Stanford
   University for Scientific Writing
University of Tsukuba
Tsukuba, Ibaraki, Japan

ACADEMIC PRESS
An imprint of Elsevier

ALZHEIMER'S DISEASE THERANOSTICS   ISBN: 978-0-12-816412-9
Copyright © 2019 Elsevier, Inc. All rights reserved.

No part of this publication may be reproduced or transmitted in any form or by any means, electronic or mechanical, including photocopying, recording, or any information storage and retrieval system, without permission in writing from the publisher. Details on how to seek permission, further information about the Publisher's permissions policies and our arrangements with organizations such as the Copyright Clearance Center and the Copyright Licensing Agency, can be found at our website: www.elsevier.com/permissions.

This book and the individual contributions contained in it are protected under copyright by the Publisher (other than as may be noted herein).

Notices

Practitioners and researchers must always rely on their own experience and knowledge in evaluating and using any information, methods, compounds or experiments described herein. Because of rapid advances in the medical sciences, in particular, independent verification of diagnoses and drug dosages should be made. To the fullest extent of the law, no responsibility is assumed by Elsevier, authors, editors or contributors for any injury and/or damage to persons or property as a matter of products liability, negligence or otherwise, or from any use or operation of any methods, products, instructions, or ideas contained in the material herein.

Publisher: Nikki Levy
Acquisition Editor: Natalie Farra
Editorial Project Manager: Tracy Tufaga
Production Project Manager: Sreejith Viswanathan
Cover Designer: Greg Harris

Academic Press is an imprint of Elsevier
125 London Wall, London EC2Y 5AS, United Kingdom
525 B Street, Suite 1650, San Diego, CA 92101, United States
50 Hampshire Street, 5th Floor, Cambridge, MA 02139, United States
The Boulevard, Langford Lane, Kidlington, Oxford OX5 1GB, United Kingdom

Working together to grow libraries in developing countries

www.elsevier.com • www.bookaid.org

This work is dedicated to my beloved parents (Magisetty Jagan Mohan and Magisetty Saraswathi), wife, and all my family members (sisters, daughter, and son). I had a praiseworthy mental support during this project and throughout my life.

Biography

M. Obulesu is a scientist in ATG Laboratories, Pune, India. He has 18 years of research and teaching experience. His research areas are multifarious, which include food science, pathology of neurodegenerative diseases such as Alzheimer's disease, and designing polymer-based biomaterials (design of hydrogels, etc.). He carried out Alzheimer's disease research and developed an aluminum-induced neurotoxicity rabbit model. His present research focuses on development of redox-active injectable hydrogels of polyion complex. His research area also includes development of metal chelators to overcome metal-induced toxicity. He is the first and corresponding author for a majority of his articles. He secured a certificate from Stanford University for a Scientific Writing course. He is on the editorial board of a few pathology journals such as the *Journal of Medical Laboratory and Diagnosis, Journal of Medical and Surgical Pathology, Annals of Retrovirals and Antiretrovirals, Kenkyu Journal of Medical Science and Clinical Research, SciFed Oncology and Cancer Research Journal,* and *Journal of Cancer and Cure.* He is also on the editorial board of nanotechnology journals such as *SciFed Nanotech Research Letters, SciFed Drug Delivery Research, Current Updates in Nanotechnology, and Journal of Nanotechnology and Materials Science.*

M. Obulesu MSc (Ph.D).
Scientist
ATG Laboratories, Pune, India
Certified by Stanford University for
Scientific Writing

Acknowledgements

I sincerely acknowledge the endeavors of my ever memorable teacher Mr. Nageswara Rao, Director, Apex Institute of English, Guntur, Andhra Pradesh, India. His worth-commending didactic skills made my English language skills reach culminating points. I also acknowledge all my teachers and mentors who significantly nurtured my research and writing skills.

Contents

CHAPTER 1

Introduction: Alzheimer's Disease Pathology and Therapeutics

ABSTRACT

Despite the extensive research for almost a century and discovery of panoply of theranostic molecules to overcome AD, there is no appropriate therapeutic strategy till date. This chapter primarily throws light on a few therapeutic interventions employed against AD which include natural compounds, natural compound conjugates, drugs, and multifarious biomaterials and discusses their pros and cons.

KEYWORDS

Alzheimer's disease; Biomaterials; Drugs; Natural compound conjugates; Natural compounds.

INTRODUCTION

Alois Alzheimer, a German psychiatrist, discovered the series of synonymous symptoms in his patient which lead to memory loss. Since then the debilitating neurodegenerative disorder is known as Alzheimer's Disease (AD). Alzheimer's Association stated that 5.4 million people in the United States are diagnosed with AD presently, and the number is expected to reach 13.8 million by 2050.[1] The alarming surge in the number of AD patients and the considerable delay in the diagnosis after the onset show the burgeoning need to streamline panoply of substantial theranostic tools at the earliest.[2] The pathophysiological events of AD entail amyloid aggregation, neurofibrillary tangle (NFT) formation, oxidative stress, apoptosis, genetic mutations, neuroinflammation, and metal dyshomeostasis[3–8].

BIOCHEMISTRY

Accumulation of extracellular amyloid protein and intracellular neurofibrillary tangles (NFTs) are the hallmarks of AD. Abnormal proteolytic cleavage of amyloid protein by β- and γ-secretases leads to the generation of Aβ plaques, which accumulate on the surface of the cell to induce neurodegeneration.[9] In addition, hyperphosphorylation of tau, a microtubular protein, also leads to the formation of intraneuronal NFTs.[5] Mounting evidence also shows that aggregation of unusually enlarged endosomes also leads to endosomal trafficking in AD.[10] Abundance of recent research also showed that metal intoxication also induces reactive oxygen species (ROS) which cause significant damage to the brain in turn leading to atrophy.[4,11] Recent postmortem studies revealed that the remarkably low copper levels induced neurodegeneration in hippocampus, entorhinal cortex, and middle-temporal gyrus, which eventually lead to the death of AD patients.[12]

Although AD stage is unpredictable by Aβ oligomers, yet their appearance at the early stage of AD attracted the researchers.[13–16] While the clearance of Aβ plays a pivotal role in AD therapeutics, excessive clearance may impair usual synaptic activity.[14,17,18] Multifactorial etiology of AD impedes the development of novel biomarkers to diagnose disease from mild to significant memory loss.[14]

THERANOSTICS

Since the discovery of multifaceted disease pathways, panoply of theranostic avenues was employed to ameliorate AD symptoms. These theranostic avenues include but are not limited to enzyme inhibitors,[19] antioxidants,[20] drugs,[21] drug delivery systems,[11] gene therapy,[22] and viral vector therapeutics.[23] Despite the remarkable protective efficacy of the AD targeted therapeutic strategies, their success in clinical trials was limited. Growing lines of evidence showed that Fyn kinase can be a better AD therapeutic target due to its initiation by a cellular prion protein and its specific interaction with tau[24]. Therefore, two hallmarks of AD can be targeted via Fyn kinase.

Antibodies

A recently designed single chain antibody fragment (scFv) known as NUsc1 showed substantial efficacy not only in detection of Aβ oligomers but also played a pivotal role in curtailing Aβ-induced oxidative stress.[25] While the employment of antibodies has been considered robust treatment against AD, their restricted stability in specific temperature range impedes their success.[26]

Alzheimer's Disease Theranostics. https://doi.org/10.1016/B978-0-12-816412-9.00001-X
Copyright © 2019 Elsevier Inc. All rights reserved.

1

Natural Compounds

Berberine, an isoquinoline alkaloid abundant in Coptidis and rhizome and Cortex phellodendri, offers copious pharmacological effects. Recent studies on neuroprotective efficacy of this natural compound showed significant amelioration of Aβ25-35-induced apoptosis.[27] In another study, Cedrin isolated from Cedrus Deodara which is abundant in the Himalayan mountain region showed substantial therapeutic effect against AD by ameliorating mitochondrial dysfunction and attenuating oxidative stress and apoptosis in pheochromocytoma (PC12) cells.[28]

Several lines of evidence showed the significant therapeutic efficacy of recently designed natural compound rhein conjugated hybrid against AD. In this study, rhein, a natural compound with adequate Aβ and tau anti-aggregating activity, was conjugated to huprine Y, an acetylcholinesterase (AChE) inhibitor. Interestingly, the hybrid showed neuroprotective efficacy by binding to catalytic anionic site and marginal anionic site of AChE, and to the catalytic site and a hitherto unknown secondary site of beta site cleaving enzyme (BACE 1).[29-32]

Resveratrol

Resveratrol (3,5,4 trihydroxy-trans stilbene), a polyphenolic natural compound, is abundant in seeds of wide range of plants such as peanuts, berries, grains, and grapes.[33] Plethora of biological activities exerted by resveratrol include antiapoptotic effects, cardioprotective properties, antitumor, antiaging, antidiabetes, antioxidant, anti-inflammatory, and neuroprotective effects.[33-35] The primary mechanism of antioxidant potential of resveratrol is through the upregulation of antioxidant enzymes such as glutathione peroxidase (GSH-Px) and superoxide dismutase (SOD).[33,36] The intricate mechanism of action of resveratrol entails a few vital factors such as sirtuin 1 (SIRT1), adenosine 5′-monophosphate activated kinase (AMPK), and nuclear factor erythroid derived 2 (Nrf2).[33,37-43]

Ginkgo Biloba

Ginkgo biloba (Ginkgoaceae), an ancient Chinese tree, leaves have extensively been used as phytomedicine and food supplement in Europe and in the United States.[33] Interestingly, ingredients of these leaves exert appreciable BBB passing efficacy and neuroprotective efficacy.[33,42-46] Mechanism of neuroprotective action of Ginkgo biloba includes pathways resulting in apoptosis.[33,45]

Drugs

Cholinesterase inhibitors such as donepezil, galantamine, and rivastigmine are extensively used in the treatment of AD currently.[46] Idalopirdine, a 5-HT$_6$ antagonist, showed adequate therapeutic efficacy in phase II clinical trials and is considered as a substantial substitute to donepezil. However, it failed to pass phase III trials.[46,47] Addition of intepirdine, another 5-HT$_6$ antagonist, to donepezil therapy showed amelioration of AD in mild to moderate AD patients.[46]

Memantine and riluzole were found to ameliorate AD symptoms by attenuating glutamine-mediated excitotoxicity.[46,48,49] A few drugs such as verubecestat and avagacestat, which showed successful Aβ clearance both in animal models and AD patients, were stopped in clinical trials due to poor efficacy or adverse effects.[46,50,51]

Although adequate number of studies have been done and in progress currently, yet there is no approved antibody-based immunotherapy to initiate Aβ clearance.[46] Because of the limited success of the Aβ therapeutics, there has been increasing interest in tau targeted therapies such as inhibition of tau phosphorylation and aggregation.[46,52] While a few drugs were effective in tau stabilization, such as Epothilone D and paclitaxel, they were discontinued due to adverse effects.[46,53]

Another etiological factor that has drawn the attention of AD researchers is neuroinflammation. Alzhemed (Tramiprosate) which was found to mitigate microglial activation was terminated in phase III clinical trial due to poor efficacy.[46,54] Although more than 200 therapeutic compounds reached phase II clinical trials in a span of more than a decade, yet no new drug was found appropriate to enter the market.[46,55,56] Novel antioxidant (selenoureido, chalcogenide) conjugated tacrine dimers exerted enhanced neuroprotective efficacy compared to tacrine itself in mouse cortical neurons, thus opening a new therapeutic avenue to combat AD.[57]

Nanotechnology

Nanoparticles (NPs) emerged as robust theranostic tools in AD treatment due to a few worth considering characteristics such as feasibility to surface functionalize with appropriate ligands, remarkable efficacy to span biological membranes, targeted delivery of therapeutic compounds, and enhanced permeation and retention ability.[14,58] Among the multifarious polymers, poly(ethylene) glycol (PEG) has been found to be biocompatible and enhance systemic circulation and is extensively used in synthesis of NPs.[14] Manifold Nps targeted against AD include Aβ-targeted Nps and Tau-targeted NPs (Fig. 1.1). Physicochemical properties of NPs such as size, charge, surface modification, composition, and shape influence the Aβ

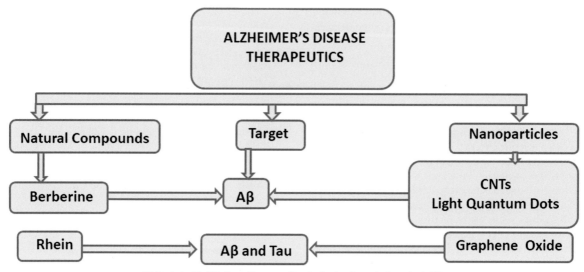

FIG. 1.1 Multifarious therapeutic strategies targeted against AD.

attenuation/provocation, thus playing an essential role in Aβ fibril formation.[14,59–64] Several lines of evidence also showed that magnetic core-plasmonic coat nanomaterial conjugated to hybrid graphene oxide with tau and Aβ facilitate specific screening of tau and Aβ in patients.[14] Another study showed that light-emitting quantum dots succeed in reaching the target site in brain and reverse Aβ-induced neuroinflammation and oxidative stress by laser therapy.[14,64]

Autophagy, a lysosomal degradative pathway, when impaired leads to neural dyshomeostasis.[65,66] To overcome this,[67] single-wall carbon nanotubes (CNTs) were designed, which ameliorated impaired autophagy and lysosomal dysfunction in glial cells of a CRND8 transgenic mouse model of AD.[14] CNTs have been found to facilitate Aβ clearance within lysosomes by attenuating the mTOR signaling pathway.[14]

An MRI nanoprobe designed recently by conjugating AβO-specific antibodies to magnetic nanostructures (SPIONs) showed significant efficacy in spanning BBB binding target.[14,68] These NPs can also facilitate the monitoring of new AD therapeutics. Georganopoulou et al., [2005] for the first time designed meticulous DNA loaded magnetic NPs which can effectively diagnose and quantify AβO through a bar code assay. According to their study, cerebrospinal fluid (CSF) concurrently interacts with both AβO antibody-conjugated magnetic particles and AβO antibody-conjugated gold NPs functionalized with double-stranded DNA. Following this interaction, a sandwich complex hosting magnetic particles and gold NPs is formed. The separation of complex was accomplished by applying magnetic field and

washing. DNA strands of the complex were separated by heating and evaluated by a bar code assay which sensitively and specifically diagnosed early-stage AD.[14]

Oxidative stress induced by ROS/reactive nitrogen species (RNS) has long been considered to play a vital role in neurodegenerative disorders in general and AD in specific. Therefore, nanotechnology employed to combat the same attracted the attention of researchers.[14,70,71] Metal dyshomeostasis also plays a crucial role in AD pathology by influencing the key enzymes involved in Aβ production and clearance.[14] With a view to overcoming the metal intoxication, metal chelation therapy has been extensively used.[72] Recently designed ultrathin graphitic phase carbon nitride (g-C_3N_4) scavenged copper (Cu^{2+}) ions and curtailed Aβ fibrillation.[73] H_2O_2-sensitive silica nanocarriers were also designed to accomplish targeted delivery of metal chelators to the H_2O_2 producing cells.[14,74]

Although multitudinous natural compounds offer appreciable therapeutic effects against AD, their bioavailability and BBB permeability impede their success considerably. To overcome these issues, recently resveratrol-loaded lipid core nanocapsules (LNCs) were designed which showed enhanced BBB permeability and amelioration of AD symptoms in rats.[75] Resveratrol-loaded LNCs were designed by interfacial deposition of the polymer.[76,77] Briefly, organic phase was prepared by dissolving trans-resveratrol, poly(ε-caprolactone), capric/caprylic triglyceride, and sorbitan monostearate in specific proportion in acetone at 40°C. Further, polysorbate was dissolved in Milli-Q water. The organic solution was added to aqueous phase on magnetic stirrer at room temperature. Finally, acetone was evaporated and contents were concentrated at 40°C to achieve LNC.[75–77]

CONCLUSION AND FUTURE PERSPECTIVES

Several novel hitherto discovered synthetic compounds not only failed to ameliorate AD symptoms but also exerted a few side effects.[33] Therefore, the current focus of AD research is on developing new bioactive compounds and delivery systems for the same to accomplish targeted delivery of theranostic compounds.[78] Although nanotechnology has been extensively used to circumvent manifold diseases, yet its success in treating neurodegenerative diseases such as AD is limited due to hitherto unpredictable issues. In addition, studies targeting multiple pathogenic factors simultaneously are in great demand.[30,79–82]

Despite the extensive review of voluminous AD data by multiple review boards, the complexity of AD pathology has been an Achilles heel that considerably limits the development of novel therapeutics.[82] One of the major challenges in formulation and targeting of NPs is limited understanding of the protein corona formed around the NPs, which considerably impedes the targeting ability.[14,60,61,84] The mainstay in current AD research is to consider challenges in available nanoparticles and enhance their efficacy to load multiple therapeutics that can target more than one etiological factor. The current unending quest for the novel therapeutic molecules or DDS may probably be of great interest to the neuroscientists and become a holy grail in the treatment of AD.

REFERENCES

1. Alzheimer's Association. Alzheimer's disease facts and figures. *Alzheimers Dement.* 2016;12:459–509.
2. Sengupta U, Portelius E, Hansson O. Tau oligomers in cerebrospinal fluid in Alzheimer's disease. *Ann Clin Transl Neurol.* 2017;4:226–235.
3. Obulesu M, Rao DM. DNA damage and impairment of DNA repair in Alzheimer's disease. *Int J Neurosci.* 2010;120:397–403.
4. Obulesu M, Venu R, Somashekhar R. Lipid peroxidation in Alzheimer's disease: emphasis on metal-mediated neurotoxicity. *Acta Neurol Scand.* 2011a;124:295–301.
5. Obulesu M, Venu R, Somashekhar R. Tau mediated neurodegeneration: an insight into Alzheimer's disease pathology. *Neurochem Res.* 2011b;36:1329–1335.
6. Obulesu M, Somashekhar R, Venu R. Genetics of Alzheimer's disease: an insight into presenilins and apolipoprotein E instigated neurodegeneration. *Int J Neurosci.* 2011c;121:229–236.
7. Obulesu M, Jhansilakshmi M. Neuroinflammation in Alzheimer's disease: an understanding of physiology and pathology. *Int J Neurosci.* 2014a;124:227–235.
8. Obulesu M, Lakshmi MJ. Apoptosis in Alzheimer's disease: an understanding of the physiology, pathology and therapeutic avenues. *Neurochem Res.* 2014b;39:2301–2312.
9. Nery LR, Silva NE, Fonseca R, Vianna MRM. Presenilin-1 targeted morpholino induces cognitive deficits, increased brain Aβ1-42 and decreased synaptic marker PSD-95 in zebrafish larvae. *Neurochem Res.* 2017;42:2959–2967.
10. Kimura N, Yanagisawa K. Traffic jam hypothesis: relationship between endocytic dysfunction and Alzheimer's disease. *Neurochem Int.* 2017;119:35–41.
11. Obulesu M, Dowlathabad MR, Raichurkar KP, Shamasundar NM. First MRI studies on aluminium maltolate treated aged New Zealand rabbit. *Psychogeriatrics.* 2016;16:263–267.
12. Xu J, Church SJ, Patassini S, et al. Evidence for widespread, severe brain copper deficiency in Alzheimer's dementia. *Metallomics.* 2017;9:1106–111.
13. Benilova I, Karran E, De Strooper B. The toxic Aβ oligomer and Alzheimer's disease: an emperor in need of clothes. *Nat Neurosci.* 2012;15:349.
14. Hajipour MJ, Santoso MR, Rezaee F, Aghaverdi H, Mahmoudi M, Perry G. Advances in Alzheimer's diagnosis and therapy: the implications of nanotechnology. *Trends Biotechnol.* 2017;35:937–953.
15. Lacor PN, Buniel MC, Chang L, et al. Synaptic targeting by Alzheimer's-related amyloid β oligomers. *J Neurosci.* 2004;24:10191–10200.
16. Nyborg AC, Moll JR, Wegrzyn RD, et al. *In vivo* and ex vivo imaging of amyloid-β cascade aggregates with a Pronucleon TM peptide. *J Alzheimers Dis.* 2013;34:957–967.
17. Cirrito JR, Yamada KA, Finn MB, et al. Synaptic activity regulates interstitial fluid amyloid-b levels *in vivo*. *Neuron.* 2005;48:913–922.
18. Parihar MS, Brewer GJ. Amyloid-b as a modulator of synaptic plasticity. *J Alzheimers Dis.* 2010;22:741–763.
19. Lazarevic-Pasti T, Leskovac A, Momic T, Petrovic S, Vasic V. Modulators of acetylcholinesterase activity: from Alzheimer's disease to anti-cancer drugs. *Curr Med Chem.* 2017;24:3283–3309.
20. Cervantes B, Ulatowski LM. Vitamin E and Alzheimer's Disease-Is It Time for Personalized Medicine? *Antioxidants (Basel).* 2017;6:pii: E45.
21. Korabecny J, Nepovimova E, Cikankova T, et al. Newly developed drugs for Alzheimer's disease in relation to energy metabolism, cholinergic and monoaminergic neurotransmission. *Neuroscience.* 2017;370:191–206.
22. Fol R, Braudeau J, Ludewig S, et al. Viral gene transfer of APPsα rescues synaptic failure in an Alzheimer's disease mouse model. *Acta Neuropathol.* 2016;131:247–266.
23. Li Y, Wang J, Zhang S, Liu Z. Neprilysin gene transfer: a promising therapeutic approach for Alzheimer's disease. *J Neurosci Res.* 2015;93:1325–1329.

24. Nygaard HB. Targeting Fyn kinase in Alzheimer's disease. *Biol Psychiatry.* 2018;83:369–376.

25. Sebollela A, Cline EN, Popova I, et al. A human scFv antibody that targets and neutralizes high molecular weight pathogenic amyloid-β oligomers. *J Neurochem.* 2017 (in press).

26. Montoliu-Gaya L, Murciano-Calles J, Martinez JC, Villegas S. Towards the improvement in stability of an anti-Aβ single-chain variable fragment, scFv-h3D6, as a way to enhance its therapeutic potential. *Amyloid.* 2017;24:167–175.

27. Liang Y, Huang M, Jiang X, Liu Q, Chang X, Guo Y. The neuroprotective effects of Berberine against amyloid β-protein-induced apoptosis in primary cultured hippocampal neurons via mitochondria-related caspase pathway. *Neurosci Lett.* 2017;655:46–53.

28. Zhao Z, Dong Z, Ming J, Liu Y. Cedrin identified from Cedrus deodara (Roxb.) G. Don protects PC12 cells against neurotoxicity induced by Aβ1-42. *Nat Prod Res.* 2017:1–4.

29. Di Pietro O, Juarez-Jimenez J, Munoz-Torrero D, Laughton CA, Luque FJ. Unveiling novel transient druggable pockets in BACE-1 through molecular simulations: conformational analysis and binding mode of multisite inhibitors. *PLoS One.* 2017;12:e0190327.

30. Perez-Areales FJ, Betari N, Viayna A, et al. Design, synthesis and multitarget biological profiling of second-generation anti-Alzheimer rhein-huprine hybrids. *Future Med Chem.* 2017;9:965–981.

31. Serrano FG, Tapia-Rojas C, Carvajal FJ, et al. Rhein-huprine derivatives reduce cognitive impairment, synaptic failure and amyloid pathology in AβPPswe/PS-1 mice of different ages. *Curr Alzheimer Res.* 2016;13:1017–1029.

32. Viayna E, Sola I, Bartolini M, et al. Synthesis and multitarget biological profiling of a novel family of rhein derivatives as disease-modifying anti-Alzheimer agents. *J Med Chem.* 2014;57:2549–2567.

33. Wasik A, Antkiewicz-Michaluk L. The mechanism of neuroprotective action of natural compounds. *Pharmacol Rep.* 2017;69:851–860.

34. Jing YH, Chen KH, Kuo PC, Pao CC, Chen JK. Neurodegeneration in streptozotocin induced diabetic rats is attenuated by treatment with resveratrol. *Neuroendocrinology.* 2013;98:116–127.

35. Kasiotis KM, Pratsinis H, Kletsas D, Haroutounian SA. Resveratrol and related stilbenes: their anti-aging and anti-angiogenic properties. *Food Chem Toxicol.* 2013;61:112–120.

36. Liu GS, Zhang ZS, Yang B, He W. Resveratrol attenuates oxidative damage and ameliorates cognitive impairment in the brain of senescence-accelerated mice. *Life Sci.* 2012;91:872–877.

37. Borra MT, Smith BC, Denu JM. Mechanism of human SIRT1 activation by resveratrol. *J Biol Chem.* 2005;280:17187–17195.

38. Chen CY, Jang JH, Li MH, Surh YJ. Resveratrol upregulates heme oxygenase-1 expression via activation of NF-E2 related factor 2 in PC12 cells. *Biochem Biophys Res Commun.* 2005;331:993–1000.

39. Dasgupta B, Milbrandt J. Resveratrol stimulates AMP kinase activity in neurons. *Proc Natl Acad Sci USA.* 2007;104:7217–7222.

40. Lopez MS, Dempsey RJ, Vemuganti R. Resveratrol neuroprotection in stroke and traumatic CNS injury. *Neurochem Int.* 2015;89:75–82.

41. Ungvari Z, Bagi Z, Feher A, et al. Resveratrol confers endothelial protection via activation of the antioxidant transcription factor Nrf2. *Am J Physiol Heart Circ Physiol.* 2010:18–24.

42. DeFeudis FV, Drieu K. Ginkgo biloba extract (EGb 761) and CNS functions: basic studies and clinical applications. *Curr Drug Targets.* 2000;1:25–58.

43. Jacobs BP, Browner WS. Ginkgo biloba: a living fossil. *Am J Med.* 2000;108:341–342.

44. Maclennan KM, Darlington CL, Smith PF. The CNS effects of Ginkgo biloba extracts and ginkolide B. *Prog Neurobiol.* 2002;67:235–257.

45. Smith JV, Luo Y. Studies on molecular mechanisms of Ginkgo biloba extract. *Appl Microbiol Biotechnol.* 2004;64:465–472.

46. Hung SY, Fu WM. Drug candidates in clinical trials for Alzheimer's disease. *J Biomed Sci.* 2017;24:47.

47. Brauser D. Two more phase 3 trials of Alzheimer's drug idalopirdine fail. *Medscape.* 2017.

48. Hunsberger HC, Weitzner DS, Rudy CC, et al. Riluzole rescues glutamate alterations, cognitive deficits, and tau pathology associated with P301L tau expression. *J Neurochem.* 2015;135:381–394.

49. Thomas SJ, Grossberg GT. Memantine: a review of studies into its safety and efficacy in treating Alzheimer's disease and other dementias. *Clin Interv Aging.* 2009;4:367–377.

50. Kennedy ME, Stamford AW, Chen X, Cox K, Cumming JN, Dockendorf MF, et al. The BACE1 inhibitor verubecestat (MK-8931) reduces CNS beta-amyloid in animal models and in Alzheimer's disease patients. *Sci Transl Med.* 2016;8:363ra150.

51. Hawkes N. Merck ends trial of potential Alzheimer's drug verubecestat. *BMJ.* 356:j845.

52. Panza F, Solfrizzi V, Seripa D, et al. Tau-centric targets and drugs in clinical development for the treatment of Alzheimer's disease. *BioMed Res Int.* 2016;2016:3245935.

53. Butler D, Bendiske J, Michaelis ML, Karanian DA, Bahr BA. Microtubule stabilizing agent prevents protein accumulation-induced loss of synaptic markers. *Eur J Pharmacol.* 2007;562:20–27.

54. Aisen PS, Gauthier S, Ferris SH, et al. Tramiprosate in mild-to-moderate Alzheimer's disease - a randomized, double-blind, placebo-controlled, multicentre study (the Alphase Study). *Arch Med Sci.* 2011;7:102–111.

55. Cummings JL, Morstorf T, Zhong K. Alzheimer's disease drug-development pipeline: few candidates, frequent failures. *Alzheimers Res Ther.* 2014;6:37.

56. Godyn J, Jonczyk J, Panek D, Malawska B. Therapeutic strategies for Alzheimer's disease in clinical trials. *Pharmacol Rep.* 2016;68:127–138.

57. Roldan-Pena JM, Alejandre-Ramos D, Lopez O, et al. New tacrine dimers with antioxidant linkers as dual drugs: anti-Alzheimer's and antiproliferative agents. *Eur J Med Chem.* 2017;138:761–773.

58. Krol S, Macrez R, Docagne F, et al. Therapeutic benefits from nanoparticles: the potential significance of nanoscience in diseases with compromise to the blood brain barrier. *Chem Rev.* 2013;113:1877–1903.

59. Cabaleiro-Lago C, Quinlan-Pluck F, Lynch I, Dawson KA, Linse S. Dual effect of amino modified polystyrene nanoparticles on amyloid b protein fibrillation. *ACS Chem Neurosci.* 2010;1:279–287.

60. Mahmoudi M, Quinlan-Pluck F, Monopoli MP, et al. Influence of the physiochemical properties of super paramagnetic iron oxide nanoparticles on amyloid b protein fibrillation in solution. *ACS Chem Neurosci.* 2013;4:475–485. 60.

61. Mahmoudi M, Monopoli MP, Rezaei M, et al. The protein corona mediates the impact of nanomaterials and slows amyloid beta fibrillation. *Chembiochem.* 2013;14:568–572.

62. Mirsadeghi S, Dinarvand R, Ghahremani MH, et al. Protein corona composition of gold nanoparticles/nanorods affects amyloid beta fibrillation process. *Nanoscale.* 2015;7:5004–5013.

63. Yoo SI, Yang M, Subramanian V, et al. Mechanism of fibrillation inhibition of amyloid peptides by inorganic nanoparticles reveal functional similarities with proteins. *Angew Chem Int Ed Engl.* 2011;50:5110.

64. Bungart BL, Dong L, Sobek D, Sun GY, Yao G, Lee JC. Nanoparticle-emitted light attenuates amyloid-b-induced superoxide and inflammation in astrocytes. *Nanomedicine.* 2014;10:15–17.

65. Moreira PI, Siedlak SL, Wang X, et al. Increased autophagic degradation of mitochondria in Alzheimer disease. *Autophagy.* 2007;3:614–615.

66. Glick D, Barth S, Macleod KF. Autophagy: cellular and molecular mechanisms. *J Pathol.* 2010;221:3–12.

67. Xue X, Wang LR, Sato Y, et al. Single-walled carbon nanotubes alleviate autophagic/lysosomal defects in primary glia from a mouse model of Alzheimer's disease. *Nano Lett.* 2014;14:5110–5117.

68. Viola KL, Sbarboro J, Sureka R, et al. Towards non-invasive diagnostic imaging of early-stage Alzheimer's disease. *Nat Nanotechnol.* 2015;10:91–98.

69. Georganopoulou DG, Chang L, Nam JM, Thaxton CS, Mufson EJ, Klein WL. Nanoparticle-based detection in cerebral spinal fluid of a soluble pathogenic biomarker for Alzheimer's disease. *Proc Natl Acad Sci USA.* 2005;102:2273–2276.

70. Andersen JK. Oxidative stress in neurodegeneration: cause or consequence? *Nat Med.* 2004;10(suppl):S18–S25.

71. Liu Y, Ai K, Ji X, et al. Comprehensive insights into the multi-anti- oxidative mechanisms of melanin nanoparticles and their application to protect brain from injury in ischemic stroke. *J Am Chem Soc.* 2017;139:856–862.

72. Sastre M, Ritchie CW, Hajji N. Metal ions in Alzheimer's disease brain. *JSM Alzheimers Dis Relat Dement.* 2015;2:1014.

73. Li M, Guan Y, Ding C, Chen Z, Ren J, Qu X. An ultrathin graphitic carbon nitride nanosheet: a novel inhibitor of metal-induced amyloid aggregation associated with Alzheimer's disease. *J Mater Chem B.* 2016;4:4072–4075.

74. Geng J, Li M, Wu L, Chen C, Qu X. Mesoporous silica nanoparticle-based H_2O_2 responsive controlled-release system used for Alzheimer's disease treatment. *Adv Healthc Mater.* 2012;1:332–336.

75. Frozza RL, Bernardi A, Hoppe JB, et al. Neuroprotective effects of resveratrol against Aβ administration in rats are improved by lipid-corenanocapsules. *Mol Neurobiol.* 2013;47:1066–1080.

76. Frozza RL, Bernardi A, Paese K, et al. Characterization of trans-resveratrol-loaded lipid-core nanocapsules and tissue distribution studies on rats. *J Biomed Nanotechnol.* 2010;6:694–703.

77. Jager E, Venturini CG, Poletto FS, et al. Sustained release from lipid-core nanocapsules by varying the core viscosity and the particle surface area. *J Biomed Nanotechnol.* 2009;5:130–140.

78. Wong HL, Wu XY, Bendayan R. Nanotechnological advances for the delivery of CNS therapeutics. *Adv Drug Deliv Rev.* 2012;64:686–700.

79. Chen Z, Digiacomo M, Tu Y, et al. Discovery of novel rivastigmine–hydroxycinnamic acid hybrids as multitargeted agents for Alzheimer's disease. *Eur J Med Chem.* 2017;125:784–792.

80. Joubert J, Foka GB, Repsold BP, Oliver DW, Kapp E, Malan SF. Synthesis and evaluation of 7-substituted coumarin derivatives as multimodal monoamine oxidase-B and cholinesterase inhibitors for the treatment of Alzheimer's disease. *Eur J Med Chem.* 2017;125:853–864.

81. Jerabek J, Uliassi E, Guidotti L, et al. Tacrine–resveratrol fused hybrids as multi-target-directed ligands against Alzheimer's disease. *Eur J Med Chem.* 2017;127:250–262.

82. Panek D, Wieckowska A, Wichur T, et al. Design, synthesis and biological evaluation of new phthalimide and saccharin derivatives with alicyclic amines targeting cholinesterases, beta-secretase and amyloid beta aggregation. *Eur J Med Chem.* 2017;125:676–695.

83. Knopman D, Alford E, Tate K, Long M, Khachaturian AS. Patients come from populations and populations contain patients. A two-stage scientific and ethics review: the next adaptation for single institutional review boards. *Alzheimers Dement.* 2017;13:940–946.

84. Mahmoudi M, Akhavan O, Ghavami M, Rezaee F, Ghiasi SM. Graphene oxide strongly inhibits amyloid beta fibrillation. *Nanoscale.* 2012;4:7322–7325.

CHAPTER 2

Early Diagnosis of Alzheimer's Disease

ABSTRACT

Early diagnosis of Alzheimer's disease (AD) is a hitherto unresolved biological riddle despite the significant progress in research. Multifarious hitherto employed diagnostic tools include development of blood and cerebrospinal fluid–based biomarkers, imaging tools such as MRI, fMRI, and PET, proteomics, etc. Despite the discovery of robust diagnostic tools, early AD diagnosis is still a mirage. To overcome this challenge a multidimensional approach that can coalesce the data from various tools may be of utmost importance to accomplish early AD diagnosis. This chapter focuses primarily on the recent developments in AD diagnosis and their merits and demerits.

KEYWORDS

Alzheimer's disease; Biomarkers; Early diagnosis; MRI; PET.

INTRODUCTION

Alzheimer's disease (AD) is a major neurodegenerative disorder with a significantly high rate of increase in patients and resulting socioeconomic burden every year.[1] However, current AD diagnosis is cumbersome, which entails medical history, clinical assessment, and physical and neurological examination, and it may be limited only to AD.[2,3] Unfortunately, autopsy is the only accurate method to corroborate AD pathology currently.[4] AD pathological molecular underpinnings involved in Aβ42 accumulation are expected to commence almost 30 years earlier than diagnosis.[5] Based on the National Institute of Aging and Alzheimer's Association updated guidelines, AD is categorized into three stages, namely preclinical, mild cognitive impairment (MCI) through AD, and dementia through AD.[3] Preclinical stage shows prior brain pathology without the symptoms of memory impairment while MCI through AD shows minor symptoms. Unlike these two stages, dementia through AD patients show memory, thinking, and behavioral symptoms with remarkable pathological aspects.[3]

Following the age of onset, AD is categorized into early-onset AD (EOAD onset <65 years) and late-onset AD (LOAD onset ≥ 65 years). Since EOAD shows higher progression rate, thorough insight into it can help in AD diagnosis and therapeutics.[6,7] Several lines of evidence suggested that nearly 60% of dementia cases are undetected at the earliest stage, and the cost associated with the patient can be significantly curtailed if detected early.[8,9] The two vital research strategies in current AD diagnosis are designing appropriate imaging biomarkers or blood biomarkers, which can be detected by an appropriate test.[10]

Among the wide range of neurochemicals studied, antioxidant glutathione (GSH)[11] and neurotransmitter gamma-aminobutyric acid (GABA)[12] measured using MEGA-PRESS[13] pulse sequence contributed significantly to the AD diagnosis.[4] More recently developed Spherical Brain Mapping (SBM) is of considerable use in discriminating AD patients from age-matched individuals without dementia symptoms.[14] It provides bidimensional maps using which MCI conversion to AD can be predicted 6 months before. Recent proteomics-based studies showed that keratin type-2 and albumin expression and white matter variations play an essential role in diagnosing MCI leading to AD.[15] The transactive response DNA binding protein 43 (TDP-43) pathology in anterior temporal pole cortex (ATPC) play a crucial role in neocortical stage of TDP-43 progression in normal aging and AD whereas spreading of TDP-43 pathology to the midfrontal cortex is a late stage found in remarkable cognitive dysfunction.[16]

CURCUMIN

Curcumin (1,7-bis-(4-hydroxy-3-methoxyphenyl)-1,6-heptadiene-3,5-dione), a popularly used culinary spice of India, shows exemplary prevention against multifarious diseases including AD. Interestingly, it has also been found to have appropriate diagnostic benefits as well.[17] Voluminous data accentuate the suitability and sensitivity of curcumin as a diagnostic molecule and invigorating nutraceutical with multi-target directive ability.[17–19]

Alzheimer's Disease Theranostics. https://doi.org/10.1016/B978-0-12-816412-9.00002-1
Copyright © 2019 Elsevier Inc. All rights reserved.

7

CREATIVE DRAWING TECHNIQUE

Surprisingly, a novel creative drawing technique called tree drawing test (TDT) has been discovered to discriminate healthy controls, early dementia of Alzheimer type (eDAT) and moderate dementia of Alzheimer type (mDAT), which accurately correlated with clinical assessment by 88%.[20] This advanced study was based on previously reported clock drawing test (CDT) but could discriminate MCI individuals successfully unlike CDT.[20-22]

Imaging Techniques

Multifarious imaging techniques currently in use for AD diagnosis include PET,[23] magnetic resonance imaging (MRI),[24] functional MRI (fMRI),[25] and magnetic resonance spectroscopy (MRS).[4,26] While the noninvasive MRI substantially supports clinician's AD diagnosis, anatomical conversion data alone are insufficient to conclude diagnosis currently.[4] Wealth of studies accentuates that early diagnosis of AD can be accomplished by multiple diagnostic approaches such as MRI scanning and metabolism images.[27] Advent of multifarious MR pulse sequence and signal processing packages/toolboxes like LCModel,[28] jMRUI,[29] Gannet,[30] and KALPANA[31] has given profound impetus to MRS in analyzing neurometabolites in various brain regions.[4]

Positron Emission Tomography Radiotracers

According to recent reports, food and drug administration (FDA) approved positron emission tomography (PET) radiotracers which help AD diagnosis significantly.[32-34] Carbon-11-labeled Pittsburgh compound B specifically identifies Aβ plaques in AD patients.[35] Despite the presence of appropriate biomarkers such as Aβ and Tau protein levels in cerebrospinal fluid (CSF), their analysis entails invasive process and limits the diagnosis.[35-37] However, significantly high cost and low availability impedes its success.

A few currently available first-generation tau tracers used in PET such as carbazole flortaucipir and the 2-arylquinolines of the THK series exhibit limitations such as poor coincidence between pathological tau load, poor sensitivity to tau in initial disease condition, and significant variation in tracer binding in cases.[38] Therefore, to overcome these challenges studies on next generation tracers are in progress currently.[38-40] Tau has also been considered as a prominent biomarker to study AD progression.[38,41,42]

GENETICS

The three pivotal genes involved in etiology of EOAD and considered as major AD biomarkers are amyloid precursor protein (APP) on chromosome 21, Presenilin 1 (PSEN1) on chromosome 14, and Presenilin 2 (PSEN2) on chromosome 1.[6] Furthermore, Apolipoprotein E (APOE) gene also plays a key role in AD pathology.[43-45] Of the 4 APOE polymorphic alleles such as ε2, ε3, and ε4, ε4 shows ~40% higher frequency in AD. In addition, APOE may also play a role in Aβ independent way by intervening with cholesterol homeostasis, vascular function, and neuroinflammation.[3,46,47] PICALM (encodes clathrin assembly lymphoid myeloid leukemia), BIN1 (encodes bridging integrator 1), CD2AP (CD2-associated protein), SORL1 (encodes sortilin related receptor), and PLD3 (encodes phospholipase D3) which are with APOE also exhibit highest risk in LOAD through enhanced Aβ generation.[5]

BIOMARKERS

Blood Biomarkers

A few hitherto unraveled blood or CSF biomarkers include (1) neuroinflammatory markers like reactive oxygen species (ROS), cytokines, chemokines, astrocytes, and activated microglia; (2) proinflammatory molecules like interleukins, interferons, tumor necrosis factors (TNFs); (3) autoantibodies; (4) trace elements like copper, zinc, iron, fatty acids; (5) fatty acids, sphingolipids, ceramides; (6) micro-RNAs; and (7) Circulating nanocomponents.[48-54] Because of the accuracy and suitability manifold recent studies showed multifarious blood biomarkers in AD diagnosis.[35,55-58]

Nabers et al. recently discovered a novel immune infrared sensor that can accurately collect all the soluble Aβ isoforms from blood plasma and detect the changing conformation of Aβ from α-helical (healthy) to pathological β sheets.[59-62] Therefore, early AD diagnosis has been made feasible. Although adequate advantages such as label-free detection, low cost, low sample volume, detection before 15–20 years of clinical symptoms support this study, yet further validation of the study is required. Furthermore, a small limitation with this study is persons without dementia may demonstrate this misfolding, and at times AD patients may not show misfolding.[61-64]

Platelets being the storehouse of pivotal proteins in par with neurons such as α-synuclein, APP, tau protein became the amenable model of bioaminergic neurons.[65-68] While the Aβ and tau are well-known CSF-based biomarkers in AD diagnosis, sample collection has been a major limitation. To overcome this challenge, growing lines of evidence showed that platelet tau measured by an indirect enzyme linked immunosorbent assay (ELISA) can also be a potential biomarker in AD diagnosis. However, the confirmation of results requires a huge number of patients, and the overlap of total tau (t-tau) with VaD are the limitations.[68–70]

Protein Biomarkers

To avoid overlapping proteins with AD pathology such as vascular and Lewy body pathologies, there is a growing need to develop novel appropriate biomarkers.[3,71] Of the 30 recently discovered protein biomarkers of AD, a membrane glycoprotein called amyloid like protein1 (APLP1) and osteopontin (SPP1) have the ability to distinguish between MCI and healthy controls.[3] Manifold APLP1 descendant proteins have been found to be substantial CSF biomarkers of Aβ generation and drug response monitoring molecules specifically for γ-secretase modulators.[3,72,73] Surprisingly, the common tryptic peptide available in all APLP1 peptides facilitates their study by selected reaction monitoring (SRM) in mass spectrometry.[3]

miRNA

Exosomal miRs such as miR-135a, -193b, and -384 play a key role in controlling the expression of amyloid precursor protein (APP), βamyloid cleaving enzyme-1 (BACE1). In line with this, recent studies showed that serum expression levels of miR-135a, miR-193b, and miR-384 analyzed using real-time quantitative reverse transcriptase polymerase chain reaction (qRT-PCR) method play a pivotal role in early AD diagnosis while miR-384 profoundly aids the categorization of AD, vascular dementia (VaD) and Parkinson's disease with dementia (PDD).[35] Several overlapping disease characteristics among AD, VaD, and PDD make AD diagnosis a herculean task.[35,74] Based on the substantial role of exosomes in Aβ degradation, it is tangible to accentuate that exosomes open a few novel therapeutic avenues in AD treatment.[35,75–77]

GLUCOSE LEVELS

Mounting evidence showed that African-Americans with diabetes show reduction of glucose level (1.3421 mg/dL) per year before the onset of AD while Caucasians show similar glucose levels.[78] Since growing number of African-American patients are at risk for diabetes, dementia diagnosis in them using glucose levels can be the appropriate strategy.[79,80] Despite the robust correlation between glucose levels and dementia in African-Americans, the underlying pathological mechanisms

are elusive. PET scans of glucose metabolism obtained by oral or intravenous administration of radiolabeled tracers were also successfully used to discriminate AD from frontotemporal dementia (FTD) and dementia of Lewy body (DLB) (Parwardhan et al., 2004).[4,81]

CONCLUSION

Since AD is a progressive neurodegenerative disorder, multidimensional diagnostic approach is essential.[4] Although a few novel protein biomarkers such as APLP1 and SPP1 showed adequate specificity in AD diagnosis, a few studies yielded contradictory results, thus leaving a potential ambiguity.[3] Despite the appropriate use of MRS in AD diagnosis, a few limitations such as extended time consumption for data procurement and the presence of corresponding hardware in clinic impede its success.[4] Staggering array of data from multidimensional diagnostic tools and involvement of artificial intelligence by a few neuroscientists is expected to lay a substantial cornerstone to streamline AD diagnosis accurately and vehemently.[4,82]

REFERENCES

1. Obulesu M, Jhansilakshmi M. Neuroprotective role of nanoparticles against Alzheimer's disease. *Curr Drug Metab.* 2016;17:142–149.
2. McKhann GM, Knopman DS, Chertkow H, et al. The diagnosis of dementia due to Alzheimer's disease: recommendations from the National Institute on Aging-Alzheimer's Association workgroups on diagnostic guidelines for Alzheimer's disease, Alzheimer's & dementia. *J Alzheimers Assoc.* 2011;7:263–269.
3. Begcevic I, Brinc D, Brown M, Martinez-Morillo E, Goldhardt O, Grimmer T. Brain-related proteins as potential CSF biomarkers of Alzheimer's disease: targeted mass spectrometry approach. *J Proteomics.* 2018;182: 12–20.
4. Mandal PK, Shukla D. Brain metabolic, structural, and behavioral pattern learning for early predictive diagnosis of alzheimer's disease. *J Alzheimers Dis.* 2018;63:935–939.
5. Guimas Almeida C, Sadat Mirfakhar F, Perdigao C, Burrinha T. Impact of late-onset Alzheimer's genetic risk factors on beta-amyloid endocytic production. *Cell Mol Life Sci.* 2018;75:2577–2589.
6. Dai. M-H, Zheng. H, Zeng. L-D, Zhang. Y. The genes associated with early-onset Alzheimer's disease. *Oncotarget.* 2018;9:15132–15143.
7. Wingo TS, Lah JJ, Levey AI, Cutler DJ. Autosomal recessive causes likely in early-onset Alzheimer disease. *Arch Neurol.* 2012;69:59–64.
8. Association As.. Alzheimer's disease facts and figures. *Alzheimers Dement.* 2018;14:367–429.

9. *Underutilization of Brain Amyloid Scans Drives Cost and Hurts Alzheimer's Disease Care [Press Release.* 2017. London, UK.

10. Gold M, Amatniek J, Carrillo MC, Cedarbaum JM, Hendrix JA, Miller BB. Digital technologies as biomarkers, clinical outcomes assessment, and recruitment tools in Alzheimer's disease clinical trials. *Alzheimers Dement (N Y).* 2018;4:234–242.

11. Saez EA, Mato Abad V, Garcia Alvarez R, et al. The quantification of Glutathione (GSH) using 1H-MRS, is possible. *Eur Soc Radiol.* 2016;1–12.

12. Limon A, Reyes-Ruiz JM, Miledi R. Loss of functional GABA(A) receptors in the Alzheimer diseased brain. *Proc Natl Acad Sci USA.* 2012;109:10071–10076.

13. Terpstra M, Henry PG, Gruetter R. Measurement of reduced glutathione (GSH) in human brain using LCModel analysis of difference-edited spectra. *Magn Reson Med.* 2003;50:19–23.

14. Martinez-Murcia FJ, Gorriz JM, Ramirez J, et al. Alzheimer's disease neuroimaging initiative. Assessing mild cognitive impairment progression using a spherical brain mapping of magnetic resonance imaging. *J Alzheimers Dis.* 2018;65:713–729.

15. Kumar A, Singh S, Verma A, Mishra VN. Proteomics based identification of differential plasma proteins and changes in white matter integrity as markers in early detection of mild cognitive impaired subjects at high risk of Alzheimer's disease. *Neurosci Lett.* 2018;S0304–3940(18)30276-3.

16. Nag S, Yu L, Boyle PA, Leurgans SE, Bennett DA, Schneider JA. TDP-43 pathology in anterior temporal pole cortex in aging and Alzheimer's disease. *Acta Neuropathol Commun.* 2018;6:33.

17. Chen M, Du ZY, Zheng X, Li DL, Zhou RP, Zhang K. Use of curcumin in diagnosis, prevention, and treatment of Alzheimer's disease. *Neural Regen Res.* 2018;13:742–752.

18. Belkacemi A, Doggui S, Dao L, Ramassamy C. Challenges associated with curcumin therapy in Alzheimer disease. *Expert Rev Mol Med.* 2011;13:e34.

19. Goozee K, Shah T, Sohrabi HR, et al. Examining the potential clinical value of curcumin in the prevention and diagnosis of Alzheimer's disease. *Br J Nutr.* 2016;115:449–465.

20. Heymann P, Gienger R, Hett A, Muller S, Laske C, Robens S, et al. Early detection of Alzheimer's disease based on the patient's creative drawing process: first results with a novel neuropsychological testing method. *J Alzheimers Dis.* 2018;63:675–687.

21. Ehreke L, Luppa M, Luck T, Wiese B, Weyerer S, Eifflaender-Gorfer S. Is the clock drawing test appropriate for screening for mild cognitive impairment? – results of the German study on ageing, cognition and dementia in primary care patients (AgeCoDe). *Dement Geriatr Cogn Disord.* 2009;28:365–372.

22. Nair AK, Gavett BE, Damman M, Dekker W, Green RC, Mandel A. Clock drawing test ratings by dementia specialists: interrater reliability and diagnostic accuracy. *J Neuropsychiatry Clin Neurosci.* 2010;22:85–92.

23. Patwardhan MB, McCrory DC, Matchar DB, Samsa GP, Rutschmann OT. Alzheimer disease: operating characteristics of PET--a meta-analysis. *Radiology.* 2004;231:73–80.

24. Vemuri P, Jack Jr CR. Role of structural MRI in Alzheimer's disease. *Alzheimer's Res Ther.* 2010;2:23.

25. Alsop DC, Press DZ. Activation and baseline changes in functional MRI studies of Alzheimer's disease. *Neurology.* 2007;69:1645–1646.

26. Mandal PK. Magnetic resonance spectroscopy (MRS) and its applications in Alzheimer's disease. *Concepts Magn Reson.* 2007;30:40–64.

27. Lu D, Popuri K, Ding GW, Balachandar R, Beg MF. Alzheimer's disease neuroimaging initiative. Multimodal and multiscale deep neural networks for the early diagnosis of Alzheimer's disease using structural MR and FDG-PET images. *Sci Rep.* 2018;8:5697.

28. Provencher SW. Automatic quantitation of localized in vivo 1H spectra with LC model. *NMR Biomed.* 2001;14:260–264.

29. Naressi A, Couturier C, Castang I, de Beer R, Graveron-Demilly D. Java-based graphical user interface for MRUI, a software package for quantitation of *in vivo*/medical magnetic resonance spectroscopy signals. *Comput Biol Med.* 2001;31:269–286.

30. Edden RA, Puts NA, Harris AD, Barker PB, Evans CJ. Gannet: a batch-processing tool for the quantitative analysis of gamma-aminobutyric acid–edited MR spectroscopy spectra. *J Magn Reson Imaging.* 2014;40:1445–1452.

31. Grewal M, Dabas A, Saharan S, Barker PB, Edden RA, Mandal PK. GABA quantitation using MEGA-PRESS: regional and hemispheric differences. *J Magn Reson Imaging.* 2016;44:1619–1623.

32. Johnson KA, Minoshima S, Bohnen NI, et al. Appropriate use criteria for amyloid PET: a report of the amyloid imaging task force, the society of nuclear medicine and molecular imaging, and the Alzheimer's association. *Alzheimers Dement.* 2013;9:e1–e16.

33. Johnson KA, Minoshima S, Bohnen NI, et al. Update on appropriate use criteria for amyloid PET imaging: dementia experts, mild cognitive impairment, and education. Amyloid imaging task force of the Alzheimer's association and society for nuclear medicine and molecular imaging. *Alzheimers Dement.* 2013b;9:e106–e109.

34. Rabinovici GD, Gatsonis C, Apgar C, et al. Impact of amyloid PET on patient management: early results from the IDEAS study. In: *Paper Presented at: Alzheimer's Association International Conference.* 2017(London, UK).

35. Yang TT, Liu CG, Gao SC, Zhang Y, Wang PC. The serum exosome derived MicroRNA-135a, -193b, and -384 were potential Alzheimer's disease biomarkers. *Biomed Environ Sci.* 2018;31:87–96.

36. Blennow K, Hampel H, Weiner M, et al. Cerebrospinal fluid and plasma biomarkers in Alzheimer disease. *Nat Rev Neurol.* 2010;6:131–144.

37. Cummings JL. Biomarkers in Alzheimer's disease drug development. *Alzheimers Dement.* 2011;7:e13–44.

38. Wren MC, Lashley T, Arstad E, Sander K. Large inter- and intra-case variability of first generation tau PET ligand binding in neurodegenerative dementias. *Acta Neuropathol Commun*. 2018;6:34.

39. Gobbi LC, Knust H, Korner M, Honer M, Czech C, Belli S. Identification of three novel radiotracers for imaging aggregated tau in Alzheimer's disease with positron emission tomography. *J Med Chem*. 2017;60: 7350–7370.

40. Hostetler ED, Walji AM, Zeng Z, Miller P, Bennacef I, Salinas C. Preclinical characterization of 18F-MK-6240, a promising PET tracer for in vivo quantification of human neurofibrillary tangles. *J Nucl Med*. 2016;57:1599–1606.

41. Arriagada PV, Growdon JH, Hedley-Whyte ET, Hyman BT. Neurofibrillary tangles but not senile plaques parallel duration and severity of Alzheimer's disease. *Neurology*. 1992;42:631–639.

42. Haroutunian V, Davies P, Vianna C, Buxbaum JD, Purohit DP. Tau protein abnormalities associated with the progression of Alzheimer disease type dementia. *Neurobiol Aging*. 2007;28:1–7.

43. Salloway S, Sperling R, Fox NC, et al. Two phase 3 trials of bapineuzumab in mild-to-moderate Alzheimer's disease. *N Engl J Med*. 2014;370:322–333.

44. Ba M, Kong M, Li X, Ng KP, Rosa-Neto P, Gauthier S. Is ApoE ε 4 a good biomarker for amyloid pathology in late onset Alzheimer's disease? *Transl Neurodegener*. 2016;5:20.

45. Egan MF, Kost J, Tariot PN, et al. Randomized trial of verubecestat for mild-to-moderate Alzheimer's disease. *N Engl J Med*. 2018;378:1691–1703.

46. Verghese PB, Castellano JM, Holtzman DM. Apolipoprotein E in Alzheimer's disease and other neurological disorders. *Lancet Neurol*. 2011;10:241–252.

47. Liu CC, Kanekiyo T, Xu H, Bu G. Apolipoprotein E and Alzheimer disease: risk, mechanisms and therapy. *Nat Rev Neurol*. 2013;9:106–118.

48. Qu BX, Gong Y, Moore C, et al. Beta-Amyloid auto-antibodies are reduced in Alzheimer's disease. *J Neuroimmunol*. 2014;274:168–173.

49. Bagyinszky E, Youn Y, An S, Kim S. Characterization of inflammatory biomarkers and candidates for diagnosis of Alzheimer's disease. *BioChip J*. 2014;8:155–162.

50. Sharma N. Exploring biomarkers for Alzheimer's disease. *J Clin Diagn Res*. 2016;10:KE01–KE06.

51. Wu J, Li L. Autoantibodies in Alzheimer's disease: potential biomarkers, pathogenic roles, and therapeutic implications. *J Biomed Res*. 2016;30:361–372.

52. Ghidoni R, Squitti R, Siotto M, Benussi L. Innovative biomarkers for alzheimer's disease: focus on the hidden disease biomarkers. *J Alzheimer's Dis*. 2018;62:1507–1518.

53. Kumar S, Reddy PH. MicroRNA-455-3p as a potential biomarker for Alzheimer's disease: an update. *Front Aging Neurosci*. 2018;10:41.

54. Kotecha AM, Correa ADC, Fisher KM. Rushworth JV olfactory dysfunction as a global biomarker for sniffing out alzheimer's disease: a meta-analysis. *Biosensors (Basel)*. 2018;8. pii: E41.

55. DeMarshall CA, Nagele EP, Sarkar A, et al. Detection of Alzheimer's disease at mild cognitive impairment and disease progression using autoantibodies as blood-based biomarkers. *Alzheimers Dement (Amst)*. 2016;3:51–62.

56. Mattsson N, Andreasson U, Zetterberg H, Blennow K, Alzheimer's Disease Neuroimaging Initiative. Association of plasma neurofilament light with neurodegeneration in patients with Alzheimer disease. *JAMA Neurol*. 2017;74:557–566.

57. Obulesu M, Jhansilakshmi M, Dhanalakshmi M, Lakshmi M. Biomarkers of Alzheimer's disease: an overview of the recent inventions. *Recent Pat Biomarkers*. 2013;3:183–187.

58. Henriksen K, O'Bryant SE, Hampel H, et al. The future of blood-based biomarkers for Alzheimer's disease. *Alzheimers Dement*. 2014;10:115–131.

59. Nabers A, Ollesch J, Schartner J, Kotting C, Genius J, Hafermann H. Amyloid-b-secondary structure distribution in cerebrospinal fluid and blood measured by an immuno-infrared-sensor: a biomarker candidate for Alzheimer's disease. *Anal Chem*. 2016a;88:2755–2762.

60. Nabers A, Ollesch J, Schartner J, Kotting C, Genius J, Haußmann U. An infrared sensor analysing label-free the secondary structure of the Abeta peptide in presence of complex fluids. *J Biophot*. 2016b;9:224–234.

61. Nabers A, Perna L, Lange J, et al. Amyloid blood biomarker detects Alzheimer's disease. *EMBO Mol Med*. 2018. pii: e8763.

62. Sarroukh R, Cerf E, Derclaye S, et al. Transformation of amyloid b(1-40) oligomers into fibrils is characterized by a major change in secondary structure. *Cell Mol Life Sci*. 2011;68:1429–1438.

63. Fiandaca MS, Zhong X, Cheema AK, et al. Plasma 24-metabolite panel predicts preclinical transition to clinical stages of Alzheimer's disease. *Front Neurol*. 2015;6:237.

64. Yang L, Rieves D, Ganley C. Brain amyloid imaging–FDA approval of florbetapir F18 injection. *N Engl J Med*. 2012;367:885–887.

65. Borroni B, Agosti C, Marcello E, Di Luca M, Padovani A. Blood cell markers in Alzheimer Disease: amyloid precursor protein form ratio in platelets. *Exp Gerontol*. 2010;45:53–56.

66. Vignini A, Sartini D, Morganti S, et al. Platelet amyloid precursor protein isoform expression in Alzheimer's disease: evidence for peripheral marker. *Int J Immunopathol Pharmacol*. 2011;24:529–534.

67. Mukaetova-Ladinska EB, Abdel-Al Z, Dodds S, et al. Platelet immunoglobulin and amyloid precursor protein (APP) as potential peripheral biomarkers for Alzheimer's disease. Findings from a pilot study. *Age Ageing*. 2012;41:408–412.

68. Mukaetova-Ladinska EB, Abdell-All Z, Andrade J, et al. Platelet tau protein as a potential peripheral biomarker in Alzheimer's disease: an explorative study. *Curr Alzheimer Res*. 2018;15:800–808.

69. Blennow K, Wallin A, Agren H, Spenger C, Siegfried J, Vanmechelen E. Tau protein in cerebrospinal fluid: a bio-

chemical marker for axonal degeneration in Alzheimer disease? *Mol Chem Neuropathol.* 1995;26:231–245.

70. Andreasen N, Minthon L, Vanmechelen E, et al. Cerebrospinal fluid tau and Abeta 42 as predictors of development of Alzheimer's disease in patients with mild cognitive impairment. *Neurosci Lett.* 1999;273:5–8.

71. Ballard C, Gauthier S, Corbett A, Brayne C, Aarsland D, Jones E. Alzheimer's disease. *Lancet.* 2011;377:1019–1031.

72. Yanagida K, Okochi M, Tagami S, et al. The 28-amino acid form of an APLP1-derived Abeta-like peptide is a surrogate marker for Abeta42 production in the central nervous system. *EMBO Mol Med.* 2009;1:223–235.

73. Sjodin S, Andersson KK, Mercken M, Zetterberg H, Borghys H, Blennow K. APLP1 as a cerebrospinal fluid biomarker for gamma-secretase modulator treatment. *Alzheimer's Res Ther.* 2015;7:77.

74. Kalaria RN. Neuropathological diagnosis of vascular cognitive impairment and vascular dementia with implications for Alzheimer's disease. *Acta Neuropathol.* 2016;131:659–685.

75. Kalani A, Tyagi A, Tyagi N. Exosomes: mediators of neurodegeneration, neuroprotection and therapeutics. *Mol Neurobiol.* 2014;49:590–600.

76. Properzi F, Ferroni E, Poleggi A, et al. The regulation of exosome function in the CNS: implications for neurodegeneration. *Swiss Med Wkly.* 2015;145:W14204.

77. van Niel G. Study of exosomes ahed new light on physiology of amyloidogenesis. *Cell Mol Neurobiol.* 2016;36:327–342.

78. Hendrie HC, Zheng M, Lane KA, et al. Changes of glucose levels precede dementia in African-Americans with diabetes but not in Caucasians. *Alzheimers Dement.* 2018;S1552–5260:30101–30108.

79. Beckles GL, Chou CF. Disparities in the prevalence of diagnosed diabetes- United States, 1999-2002 and 2011-2014. *MMWR Morb Mortal Wkly Rep.* 2016;65:1265–1269.

80. Whitlow CT, Sink KM, Divers J, et al. Effects of type 2 diabetes on brain structure and cognitive function: African American-Diabetes Heart Study MIND. *AJNR Am J Neuroradiol.* 2015;36:1648–1653.

81. Marcus C, Mena E, Subramaniam RM. Brain PET in the diagnosis of Alzheimer's disease. *Clin Nucl Med.* 2014;39:e413–e422.

82. Shen D, Wu G, Suk HI. Deep learning in medical image analysis. *Annu Rev Biomed Eng.* 2017;19:221–248.

CHAPTER 3

Antioxidants in Alzheimer's Therapy

ABSTRACT

Antioxidants have long been appreciable therapeutic arsenals in Alzheimer's disease (AD) treatment. Despite a few vital roles of reactive oxygen species in physiology, they contribute to oxidative stress when overproduced due to an impairment of cellular antioxidants. Since oxidative stress plays a crucial role in pathology of manifold diseases, antioxidant therapeutics has been gaining ground in current research. This chapter primarily focuses on multifarious antioxidants and their neuroprotective efficacy against AD. It also focuses on a few novel methods used to enhance the antioxidants' efficacy significantly. A few antioxidants when amalgamated with other therapeutics such as enzyme inhibitors showed remarkably high therapeutic efficacy against AD. However, there is a growing need to develop more appropriate molecules or hybrids to enhance AD therapy. A few nanotechnological maneuvers showed significant therapeutic efficacy are also included.

KEYWORDS

Alzheimer's disease; Antioxidants; Lipoic acid; Nanotechnology; Neuroinflammation.

INTRODUCTION

Aging leads to enhanced production of reactive oxygen species (ROS) and steady deterioration of antioxidant systems, which contributes remarkably for oxidative stress. To encounter oxidative stress in Alzheimer's disease (AD) and type 2 diabetes (T2D), robust antioxidant systems are suitable therapeutic approaches.[1,2] Drugs used in AD and T2D treatment exerted greater risk of side effects, and to overcome these issues studies were focused on herbal medicines.[2,3] Despite the copious studies and voluminous documents in Indian Ayurveda for the treatment of T2D and AD, the molecular scientific basis is elusive.[2,4] It has been found that T2D enhances AD risk by 2 folds.[2,5]

Mitochondrial dysfunction and deteriorated antioxidant capacity influenced by AMPK/SIRT/PGC-1α signaling were found to be involved in neurodegeneration in the cortex due to diabetes.[6] Growing evidence shows that Aβ accumulation and oxidative stress in mitochondria followed by energy depletion in early AD are the associated events. In addition, enhanced oxidative stress also was found to increase Aβ aggregation and tau hyperphosphorylation.[7–9]

Selenium, a vital mineral with ample physiological roles such as growth and function of cells, plays a pivotal role in combating free radical instigated cell damage.[9,10] Ebselen [2-phenyl-1,2-benzisoselenazol-3(2H)-one] that acts like glutathione peroxidase reduces peroxides and offers appreciable protection to cells.[9,11] Interestingly, it also exhibits anti-inflammatory effect and attenuates iron provoked tau phosphorylation in AD treatment.[9,12] Amalgamation of Ebselen and donepezil opened the new avenues for the development of selenpezil, which exerts cumulative therapeutic benefits and no toxicity or mortality in mice at high doses as well.[9,13]

ANTIOXIDANTS

Antioxidants have long been substantial and amenable therapeutic arsenals for multifarious diseases such as AD and cancer. Significantly high oxygen (20% of uptaken oxygen) has been consumed by central nervous system (CNS) despite its only 2% contribution to body weight.[6,14] The major reasons for the vulnerability of brain to oxidative stress are utilization of 25% of total energy and remarkably low antioxidant enzyme potential.[6,15]

Plant Extracts

Plants are the richest source of pivotal therapeutic compounds that significantly contribute to human health in manifold ways.[16] Despite the discovery and extensive research conducted on these compounds, there are several compounds to be discovered and utilized.[16] Neferine, a bisbenzylisoquinoline alkaloid constituent of lotus (*Nelumbo nucifera*) plant which plays an essential role in multifarious disease therapeutics, offers significant cytoprotective efficacy in the treatment of AD.[16] It has been found to show neuroprotective efficacy through autophagy.[16,17] To detect neferine as the potential inhibitory compound of butyrylcholinesterase, high performance liquid chromatography (HPLC)

Alzheimer's Disease Theranostics. https://doi.org/10.1016/B978-0-12-816412-9.00003-3
Copyright © 2019 Elsevier Inc. All rights reserved.

ultraviolet analysis assisted electron spray ionization ion trap time of flight spectrometry was employed.[16,18] In another study, embryo extract of *N. nucifera* exhibited significant inhibitory effect on β-site amyloid precursor protein cleaving enzyme-1 (BACE-1), acetylcholine esterase (AChE), and butyrylcholinesterase (BChE).[16,19] Neferine nanoparticles also showed enhanced efficacy of neferine to cure diseases like malaria.[16,20]

Panoply of data reveals that robust antioxidant, antiglucosidase, and anticholinesterase activities of methanolic extracts and chloroform fractions of *Buchanania axillaris*, *Hemidesmus indicus*, and *Rus mysorensis* significantly ameliorate AD and T2D pathology.[2] Mounting evidence suggests a substantial pathological link between T2D pathology and AD and T2D doubles the AD risk.[2,5]

Withanolides

Withania sominifera, an Indian ginseng or ashwagandha extract which has long been used to ameliorate memory, showed pronounced neuroprotective effects against Aβ and acrolein toxicity in SK-N-SH cells.[21,22] Remarkable reduction of ROS and suppression of acetylcholinesterase activity has been found to be the underlying cause of protection.[22] Withanolides, the multifarious biochemical ingredients such as alkaloids, steroidal lactones, and saponins, showed protective efficacy against Aβ and H_2O_2 toxicity.[22–25]

NEUROINFLAMMATION-TARGETED THERAPEUTICS

Neuroinflammation has long been considered the foremost pathological event which if combated successfully can reverse AD symptoms.[26,27] Mounting evidence suggested that α-linolenic acid (ALA) supplemented diet mitigated the oxidative stress and inflammation by perturbing nuclear factor-kappa B signaling pathway.[27,28] Additionally, it has also been shown that omega-3 fatty acids like ALA suppressed the ROS and RNS production in macrophage cells.[27,29]

Interleukin 6, interleukin 1β, and TNF-α were found to promote Aβ-induced pathology. Their increased production also impedes Aβ scavenging, specifically in glial cells.[27,30,31] Perilla oil enriched with ALA ameliorated cognitive function in $A\beta_{25-35}$ treated mouse model of AD.[27,32,33] ALA significantly curtailed $A\beta_{25-35}$ provoked production of nitric oxide and proinflammatory cytokines such as interleukin 6 (IL6) and tumor necrosis factor-α (TNF-α) in C6 glial cells.[27] Furthermore, ALA attenuated the ROS generation by augmenting the nuclear factor-erythroid 2 related factor (Nrf-2) and provoking heme-oxygenase-1 (HO-1) in C6 glial cells.[27] ALA also exhibits antioxidant activity by decreasing lipid peroxidation and regulating antioxidant enzymes superoxide dismutase, glutathione peroxidase, and catalase.[27] Being a robust antioxidant and anti-inflammatory agent, Ferulic acid has extensively been employed as a scaffold to target AD by tacrine-ferulic acid hybrids.[9,35–38]

Antioxidant Enzymes

Low levels of ROS play a pivotal role in physiological events such as cellular signaling, pro-survival pathways, or instigation of transcription factors mediating cellular response.[9,39] An array of ROS comprises oxygen obtained very reactive and short-lived molecules.[9] Of these, superoxide, hydroxyl radical, and hydrogen peroxide exert cytotoxic effect.[9,40] Plethora of biological components entailed in ROS generation includes mitochondria, NADPH oxidases, xanthine oxidase, peroxisomes, or endoplasmatic reticulum.[9,41] High oxygen utilization, presence of redox-active metals like iron or copper, and increased levels of polyunsaturated fatty acids render brain vulnerable to oxidative stress.[9,42] Antioxidant enzymes such as superoxide dismutase, glutathione peroxidase, and catalase play an essential role in curtailing ROS levels.[9,43]

Limitations

Although antioxidants have been biocompatible, amenable, and substantial therapeutic molecules, they showed low potential in AD treatment. A few reasons ascertained are inadequate dose, inadequate therapy interval, other potential causes than oxidative stress, insufficiency of single antioxidant's potential to combat oxidative stress, and introduction of antioxidant therapy in advanced stages of AD.[9,44–46]

MELATONIN

Mounting evidence suggests that melatonin (*N*-acetyl-5-methoxytryptamine) which plays a role in circadian rhythm also mitigates ROS and RNS production and provokes the antioxidant enzyme activity.[9,47] Additionally, melatonin can deteriorate Aβ load, tau hyperphosphorylation, and kainic acid–induced microglial and astroglial reactions.[9,48–50] With these appreciable therapeutic benefits, melatonin has extensively been used as an exemplary scaffold to incorporate multifarious compounds such as melatonin donepezil hybrids.[9,51] Studies on mouse model showed considerable protective efficacy of melatonin against scopolamine-induced myelin basic protein reduction by enhancing the brain-derived neurotrophic factor (BDNF) and tropomyosin receptor kinase B (TrkB) in the mouse dentate gyrus.[52]

LIPOIC ACID

Lipoic acid is another potential antioxidant that can curb ROS generation by chelating redox-active transition metals augmenting reduced glutathione levels.[9,53] Moreover, several lines of evidence also showed that it is actively involved in restoring a few biogenic antioxidants.[9,54] In line with this, lipoic acid has been accepted and validated as an amenable scaffold for incorporating bioactive compounds.[9]

NANOTECHNOLOGY

Despite the remarkable therapeutic properties of resveratrol against AD such as anti-inflammatory, antioxidative, and antiamyloid aggregation activity, low bioavailability hinders its success. To overcome this issue 28 novel mock stilbene compounds have been synthesized, and their efficacy was evaluated in BV2 microglia cell lines.[55-58] Wealth of studies has shown that tacrine-phenolic acid dihybrids and tacrine-phenolic acid-ligustrazine trihybrids significantly enhances the protective efficacy against AD by cholinesterase inhibitory activity and antiamyloid aggregation activity.[59-66]

Ligustrazine (2,3,5,6-tetramethylpyrazine, TMP), a vital alkaloid in traditional Chinese medicine Rhizoma Chuanxiong (Ligusticum chuanxiong Hort.), exerts therapeutic benefits such as antioxidation and anti-inflammatory effects in turn showing neuroprotection.[66] In addition, similar neuroprotective properties were found in phenolic acids such as caffeic acid, salvianolic acid, p-coumaric acid, and ferulic acid.[66-73] Therefore, ligustrazine was coupled with phenolic acids which showed profound improvement in the neuroprotective efficacy through pronounced cholinesterase inhibition compared to ligustrazine alone.[66,74-76]

Wealth of studies has also shown that micelle or liposome enclosed omega-3 fatty acids like ALA substantially reduce hydroxyl compounds released from oxidation products and offer protection from oxidative damage.[27,77] Furthermore, omega-3 fatty acids can be encapsulated into membrane phospholipids or attached to membrane receptors and used to reduce ROS generation.[27,78]

CONCLUSION

Multifarious drugs currently available offer mere short-term assistance but fail to target the actual cause of AD.[9,79] Combating oxidative stress by decreasing ROS/RNS levels and chelating metals are a few possible oxidative stress-releasing events. Natural antioxidants such as melatonin, lipoic acid, andferulic acid with high antioxidant efficacy were amalgamated with donepezil also showed appreciable therapeutic effect. Employing a few donepezil scaffolds in association with antioxidant molecules like lipoic acid and ferulic acid yielded success. Although extensive research has been done on antioxidants and their efficacy has been well proven, further robust studies are still required to enhance their antioxidative ability.

REFERENCES

1. Mittal K, Katare DP. Shared links between type 2 diabetes mellitus and Alzheimer's disease: a review, diabetes and metabolic syndrome. *Clin Res Rev*. 2016;10:S144–S149.
2. Penumala M, Zinka RB, Shaik JB, Mallepalli SKR, Vadde R, Amooru DG. Phytochemical profiling and *in vitro* screening for anticholinesterase, antioxidant, antiglucosidase and neuroprotective effect of three traditional medicinal plants for Alzheimer's Disease and Diabetes Mellitus dual therapy. *BMC Complement Altern Med*. 2018;18:77.
3. Tomomi F, Hiroshi I, Kaori I, Yumiko K, Toshio O. Effect of two α-glucosidase inhibitors, voglibose and acarbose, on postprandial hyperglycemia correlates with subjective abdominal symptoms. *Metabolism*. 2005;54:387–390.
4. Rao RV, Descamps O, John V, Bredesen DE. Ayurvedic medicinal plants a review. *Alzheimers Res Ther*. 2012;4:22.
5. Giuseppe V, Stephanie JF, Ralph NM. The role of type 2 diabetes in neurodegeneration. *Neurobiol Dis*. 2015;84:22–38.
6. Roy Chowdhury S, Djordjevic J, Thomson E, Smith DR, Albensi BC, Fernyhough P. Depressed mitochondrial function and electron transport Complex II-mediated H2O2 production in the cortex of type 1 diabetic Rodents. *Mol Cell Neurosci*. 2018;90:49–59.
7. Radi E, Formichi P, Battisti C, Federico A. Apoptosis and oxidative stress in neurodegenerative diseases. *J Alzheimers Dis*. 2014;42:S125–S152.
8. Zhao Y, Zhao B. Oxidative stress and the pathogenesis of Alzheimer's disease. *Oxid Med Cell Longev*. 2013;2013:316523.
9. Mezeiova E, Spilovska K, Nepovimova E, et al. Profiling donepezil template into multipotent hybrids with antioxidant properties. *J Enzyme Inhib Med Chem*. 2018;33:583–606.
10. Allmang C, Wurth L, Krol A. The selenium to selenoprotein pathway in eukaryotes: more molecular partners than anticipated. *Biochim Biophys Acta*. 2009;1790:1415–1423.
11. Wilson SR, Zucker PA, Huang RRC, Spector A. Development of synthetic compounds with glutathione peroxidase activity. *J Am Chem Soc*. 1989;111:5936–5939.
12. Xie L, Zheng W, Xin N, et al. Ebselen inhibits iron-induced tau phosphorylation by attenuating DMT1 up-regulation and cellular iron uptake. *Neurochem Int*. 2012;61:334–340.

13. Luo Z, Sheng J, Sun Y, et al. Synthesis and evaluation of multi-target-directed ligands against Alzheimer's disease based on the fusion of donepezil and ebselen. *J Med Chem.* 2013;56:9089–9099.
14. Foster KA, Galeffi F, Gerich FJ, Turner DA, Muller M. Optical and pharmacological tools to investigate the role of mitochondria during oxidative stress and neurodegeneration. *Prog Neurobiol.* 2006;79:136–171.
15. Halliwell B. Oxidative stress and neurodegeneration: where are we now? *J Neurochem.* 2006;97:1634–1658.
16. Marthandam Asokan S, Mariappan R, Muthusamy S, Velmurugan BK. Pharmacological benefits of neferine - a comprehensive review. *Life Sci.* 2018;199:60–70.
17. Coe BJ, Fielden J, Foxon SP, et al. Combining very large quadratic and cubic nonlinear optical responses in extended, tris-chelate metallochromophores with six pi-conjugated pyridinium substituents. *J Am Chem Soc.* 2010;132(2010):3496–3513.
18. Lin Z, Wang H, Fu Q, et al. Simultaneous separation, identification and activity evaluation of three butyryl-cholinesterase inhibitors from *Plumula nelumbinis* using on-line HPLC-UV coupled with ESI-IT-TOF-MS and BChE biochemical detection. *Talanta.* 2013;110:180–189.
19. Jung HA, Karki S, Kim JH, Choi JS. BACE1 and cholinesterase inhibitory activities of *Nelumbo nucifera* embryos. *Arch Pharm Res.* 2015;38:1178–1187.
20. Santhoshkumar T, Rahuman AA, Rajakumar G, et al. Synthesis of silver nanoparticles using *Nelumbo nucifera* leaf extract and its larvicidal activity against malaria and filariasis vectors. *Parasitol Res.* 2011;108:693–702.
21. Ven Murthy MR, Ranjekar PK, Ramassamy C, et al. Scientific basis for the use of Indian Ayurvedic medicinal plants in the treatment of neurodegenerative disorders: ashwagandha. *Cent Nerv Sys Agents Med Chem.* 2010;10:238–246.
22. Singh M, Ramassamy C. *In vitro* screening of neuroprotective activity of Indian medicinal plant *Withania somnifera*. *J Nutr Sci.* 2017;6:1–5.e54.
23. Kuboyama T, Tohda C, Komatsu K. Withanoside IV and its active metabolite, sominone, attenuate Aβ(25–35)-induced neurodegeneration. *Eur J Neurosci.* 2006;23:1417–1426.
24. Jayaprakasam B, Padmanabhan K, Nair MG. Withanamides in Withania somnifera fruit protect PC-12 cells from β-amyloid responsible for Alzheimer's disease. *Phytother Res.* 2010;24:859–863.
25. Kumar S, Seal CJ, Howes MJ, et al. *In vitro* protective effects of *Withania somnifera* (L.) dunal root extract against hydrogen peroxide and β-amyloid(1–42)-induced cytotoxicity in differentiated PC12 cells. *Phytother Res.* 2010;24:1567–1574.
26. McGeer PL, Itagaki S, Tago H, McGeer EG. Reactive microglia in patients with senile dementia of the Alzheimer's type are positive for the histocompatibility glycoprotein HLA-DR. *Neurosci Lett.* 1987;79:195–200.
27. Lee AY, Lee MH, Lee S, Cho EJ. Neuroprotective effect of alpha-linolenic acid against Aβ-mediated inflammatory responses in C6 glial cell. *J Agric Food Chem.* 2018;66:4853–4861.
28. Ren J, Chung SH. Anti-inflammatory effect of α-linolenic acid and its mode of action through the inhibition of nitric oxide production and inducible nitric oxide synthase gene expression via NF- κB and mitogen-activated protein kinase pathways. *J Agric Food Chem.* 2007;55:5073–5080.
29. Ambrozova G, Pekarova M, Lojek A. Effect of polyunsaturated fatty acids on the reactive oxygen and nitrogen species production by raw 264.7 macrophages. *Eur J Nutr.* 2010;49:133.
30. Tachida Y, Nakagawa K, Saito T, Saido TC, Honda T, Saito Y. Interleukin-1 beta up-regulates TACE to enhance alpha-cleavage of APP in neurons: resulting decrease in Aβ production. *J Neurochem.* 2008;104:1387–1393.
31. Solito E, Sastre M. Microglia function in Alzheimer's disease. *Front Pharmacol.* 2012;3:14.
32. Lee AY, Choi JM, Lee J, Lee MH, Lee S, Cho EJ. Effects of vegetable oils with different fatty acid compositions on cognition and memory ability in Aβ25–35-induced Alzheimer's disease mouse model. *J Med Food.* 2016;19:912–921.
33. Lee AY, Lee MH, Lee S, Cho EJ. Alpha-linolenic acid from Perilla f rutescens var. Japonica oil protects Aβ-induced cognitive impairment through regulation of APP processing and Aβ degradation. *J Agric Food Chem.* 2017;65:10719–10729.
34. Pal M, Ghosh M. Studies on comparative efficacy of α-linolenic acid and α-eleostearic acid on prevention of organic mercury induced oxidative stress in kidney and liver of rat. *Food Chem Toxicol.* 2012;50:1066–1072.
35. Fang L, Kraus B, Lehmann J, et al. Design and synthesis of tacrine–ferulic acid hybrids as multi-potent anti-Alzheimer drug candidates. *Bioorg Med Chem Lett.* 2008;18:2905–2909.
36. Sgarbossa A, Giacomazza D, di Carlo M. Ferulic acid: a hope for alzheimer's disease therapy from plants. *Nutrients.* 2015;7:5764–5782.
37. Benchekroun M, Bartolini M, Egea J, et al. Novel tacrine grafted Ugi adducts as multipotent anti-Alzheimer drugs: a synthetic renewal in tacrine-ferulic acid hybrids. *ChemMedChem.* 2015;10:523–539.
38. Chen Y, Sun J, Fang L, et al. Tacrine–ferulic acid–nitric oxide (NO) donor trihybrids as potent, multifunctional acetyl- and butyrylcholinesterase inhibitors. *J Med Chem.* 2012;55:4309–4321.
39. Patten DA, Germain M, Kelly MA, Slack RS. Reactive oxygen species: stuck in the middle of neurodegeneration. *J Alzheimers Dis.* 2010;20(suppl 2):S357–S367.
40. Bolisetty S, Jaimes EA. Mitochondria and reactive oxygen species: physiology and pathophysiology. *Int J Mol Sci.* 2013;14:6306–6344.
41. Kim GH, Kim JE, Rhie SJ, Yoon S. The role of oxidative stress in neurodegenerative diseases. *Exp Neurobiol.* 2015;24:325–340.
42. Wang X, Michaelis EK. Selective neuronal vulnerability to oxidative stress in the brain. *Front Aging Neurosci.* 2010;2:12.

43. Dasuri K, Zhang L, Keller JN. Oxidative stress, neurode-generation, and the balance of protein degradation and protein synthesis. *Free Radic Biol Med.* 2013;62:170–185.

44. Pratico D. Oxidative stress hypothesis in Alzheimer's disease: a reappraisal. *Trends Pharmacol Sci.* 2008;29:609–615.

45. Firuzi O, Miri R, Tavakkoli M, Saso L. Antioxidant therapy: current status and future prospects. *Curr Med Chem.* 2011;18:3871–3888.

46. Murphy MP. Antioxidants as therapies: can we improve on nature? *Free Radic Biol Med.* 2014;66:20–23.

47. Miller E, Morel A, Saso L, Saluk J. Melatonin redox activity. Its potential clinical applications in neurodegenerative disorders. *Curr Top Med Chem.* 2015;15:163–169.

48. Lahiri DK. Melatonin affects the metabolism of the beta amyloid precursor protein in different cell types. *J Pineal Res.* 1999;26:137–146.

49. Rosales-Corral S, Tan DX, Reiter RJ, et al. Orally administered melatonin reduces oxidative stress and proinflammatory cytokines induced by amyloid-beta peptide in rat brain: a comparative, *in vivo* study versus vitamin C and E. *J Pineal Res.* 2003;35:80–84.

50. Li XC, Wang ZF, Zhang JX, Wang Q, Wang JZ. Effect of melatonin on calyculin A-induced tau hyperphosphorylation. *Eur J Pharmacol.* 2005;510:25–30.

51. Ramos E, Egea J, de Los, Rios C, Marco-Contelles J, Romero A. Melatonin as a versatile molecule to design novel multitarget hybrids against neurodegeneration. *Future Med Chem.* 2017;9:765–780.

52. Chen BH, Park JH, Lee TK, Song M, Kim H, Lee JC. Melatonin attenuates scopolamine-induced cognitive impairment via protecting against demyelination through BDNF-TrkB signaling in the mouse dentate gyrus. *Chem Biol Interact.* 2018;285:8–13.

53. Di Domenico F, Barone E, Perluigi M, Butterfield DA. Strategy to reduce free radical species in Alzheimer's disease: an update of selected antioxidants. *Expert Rev Neurother.* 2015;15:19–40.

54. Castaneda-Arriaga R, Alvarez-Idaboy JR. Lipoic acid and dihydrolipoic acid. a comprehensive theoretical study of their antioxidant activity supported by available experimental kinetic data. *J Chem Inf Model.* 2014;54:1642–1652.

55. Meng Y, Ma QY, Kou XP, et al. Effect of resveratrol on activation of nuclear factor kappa-B and inflammatory factors in rat model of acute pancreatitis. *World J Gastroenterol.* 2005;11:525–528.

56. Bournival J, Quessy P, Martinoli MG. Protective effects of resveratrol and quercetin against MPP┼-induced oxidative stress act by modulating markers of apoptotic death in dopaminergic neurons. *Cell Mol Neurobiol.* 2009;29:1169–1180.

57. Ladiwala AR, Lin JC, Bale SS, et al. Resveratrol selectively remodels soluble oligomers and fibrils of amyloid Abeta into off-pathway conformers. *J Biol Chem.* 2010;285:24228–24237.

58. Fang Y, Xia W, Cheng B, et al. Design, synthesis, and biological evaluation of compounds with a new scaffold as anti-neuroinflammatory agents for the treatment of Alzheimer's disease. *Eur J Med Chem.* 2018;149(2018):129–138.

59. So EC, Wong KL, Huang TC, Tasi SC, Liu CF. Tetramethylpyrazine protects mice against thioacetamide-induced acute hepatotoxicity. *J Biomed Sci.* 2002;9:410–414.

60. Zhang C, Wang SZ, Zuo PP, Cui X, Cai J. Protective effect of tetramethylpyrazine on learning and memory function in D-galactose-lesioned mice. *Chin Med Sci J.* 2004;19:180–184.

61. Kao TK, Ou YC, Kuo JS, et al. Neuroprotection by tetramethylpyrazine against ischemic brain injury in rats. *Neurochem Int.* 2006;48:166–176.

62. Chen H, Li G, Zhan P, Liu X. Ligustrazine derivatives. Part 5: design, synthesis and biological evaluation of novel ligustrazinyloxy-cinnamic acid derivatives as potent cardiovascular agents. *Eur J Med Chem.* 2011;46:5609–5615.

63. Kim M, Kim SO, Lee M, et al. Tetramethylpyrazine, a natural alkaloid, attenuates proinflammatory mediators induced by amyloid beta and interferon-gamma in rat brain microglia. *Eur J Pharmacol.* 2014;740:504–511.

64. Lu C, Zhang J, Shi X, et al. Neuroprotective effects of tetramethylpyrazine against dopaminergic neuron injury in a rat model of Parkinson's disease induced by MPTP. *Int J Biol Sci.* 2014;10:350–357.

65. Zhang T, Gu J, Wu L, et al. Neuroprotective and axonal outgrowth-promoting effects of tetramethylpyrazine nitrone in chronic cerebral hypoperfusion rats and primary hippocampal neurons exposed to hypoxia. *Neuropharmacology.* 2017;118:137–147.

66. Li G, Hong G, Li X, et al. Synthesis and activity towards Alzheimer's disease *in vitro*: tacrine, phenolic acid and ligustrazine hybrids. *Eur J Med Chem.* 2018;148:238–254.

67. Picone P, Nuzzo D, Di Carlo M. Ferulic acid: a natural antioxidant against oxidative stress induced by oligomeric A-beta on sea urchin embryo. *Biol Bull.* 2013;224:18–28.

68. Huang Y, Jin M, Pi R, et al. Protective effects of caffeic acid and caffeic acid phenethyl ester against acrolein-induced neurotoxicity in HT22 mouse hippocampal cells. *Neurosci Lett.* 2013;535:146–151.

69. Qu W, Huang H, Li K, Qin C. Danshensu-mediated protective effect against hepatic fibrosis induced by carbon tetrachloride in rats. *Pathol Biol (Paris).* 2014;62:348–353.

70. Park SY, Ahn G, Um JH, et al. Hepatoprotective effect of chitosan-caffeic acid conjugate against ethanol-treated mice. *Exp Toxicol Pathol.* 2017;69:618–624.

71. Ekinci Akdemir FN, Albayrak M, Calik M, Bayir Y, Gulcin I. The protective effects of p-coumaric acid on acute liver and kidney damages induced by cisplatin. *Biomedicines.* 2017;5:E18.

72. Abdel-Moneim A, Yousef AI, Abd El-Twab SM, Abdel Reheim ES, Ashour MB. Gallic acid and p-coumaric acid attenuate type 2 diabetes induced neurodegeneration in rats. *Metab Brain Dis.* 2017;32:1279–1286.

73. Wang Y, Zhang X, Xu C, Zhang G, Zhang Z, Yu P. Synthesis and biological evaluation of danshensu and tetramethylpyrazine conjugates as cardioprotective agents. *Chem Pharm Bull (Tokyo)*. 2017;65:381–388.

74. Wang P, Zhang H, Chu F, Xu X, Lin J, Chen C. Synthesis and protective effect of new ligustrazine-benzoic acid derivatives against CoCl2-induced neurotoxicity in differentiated PC12 cells. *Molecules*. 2013;18:13027–13042.

75. Li G, Xu X, Xu K, et al. Ligustrazinyl amides: a novel class of ligustrazine-phenolic acid derivatives with neuroprotective effects. *Chem Cent J*. 2015a;9:9.

76. Li G, Tian Y, Zhang Y, et al. A novel ligustrazine derivative T-VA prevents neurotoxicity in differentiated PC12 cells and protects the brain against ischemia injury in MCAO rats. *Int J Mol Sci*. 2015b;16:21759–21774.

77. Miyashita K. Paradox of omega-3 PUFA oxidation. *Eur J Lipid Sci Technol*. 2014;116:1268–1279.

78. Serini S, Fasano E, Piccioni E, Cittadini ARM, Calviello G. Dietary n-3 polyunsaturated fatty acids and the paradox of their health benefits and potential harmful effects. *Chem Res Toxicol*. 2011;24:2093–2105.

79. Zemek F, Drtinova L, Nepovimova E, et al. Outcomes of Alzheimer's disease therapy with acetylcholinesterase inhibitors and memantine. *Expert Opin Drug Saf*. 2014;13:759–774.

CHAPTER 4

Natural Products in the Treatment of Alzheimer's Disease

ABSTRACT

Natural products have been used since ages in the treatment of multifarious diseases. This chapter primarily highlights the recently employed natural products in the treatment of Alzheimer's disease. The primary advantage with the use of natural products in the treatment is to nullify the toxic side effects which are usually induced by multifarious drug molecules currently. The profound therapeutic efficacy of curcumin, chitosan, resveratrol, and epigallocatechin 3-gallate has been described in this chapter.

KEYWORDS

Alzheimer's disease; Curcumin; Drug delivery systems; Flavonoids; Natural products.

INTRODUCTION

Wealth of studies has accentuated that curtailing reactive oxygen species (ROS) and Aβ production are the targeted strategies and the mainstay of current Alzheimer's disease (AD) research. Natural products emerge as extraordinary therapeutic agents with copious benefits such as enhanced bioavailability and low toxicity.[1] Multifarious natural products extracted from sources such as plants, animals, and microorganisms include alkaloids, carbohydrates, glycosides, terpenoids, steroids, and phenolics.[1] Natural product–based drugs have been gaining ground in medicine since ages.[2–4]

Indian Ayurveda and traditional Chinese medicine (TCM) have been followed throughout the world currently.[5–7] Interestingly, they also opened several innovative avenues in the development of pharmacological products by using plant extracts or small molecules with adequate molecular mechanisms.[6,8,9] The galanthamine alkaloid is extracted from *Galanthus woronowii* and belongs to Amaryllidaceae enhanced acetylcholine levels by inhibiting acetylcholinesterase in AD treatment.[10–13,14]

FLAVONOIDS

Plethora of flavonoids were found to span blood-brain barrier (BBB) effectively, thus playing a vital role in the treatment of neurodegenerative diseases such as AD and Parkinson's disease (PD).[6,15] Flavonoids are abundantly available in plants, fruits, and vegetables and offer manifold pharmacological and therapeutic effects.[6,16,17] Therefore, they are part of human diet. Approximately 9000 flavonoids have been discovered until now and are placed in six subclasses depending on their structure. They are flavonols (rutin, quercetin), flavanols (catechin, epicatechin, and epigallocatechin), isoflavones (genistein, daidzein, glycetin, and formanantine), anthocyanidins (cyanidin, malvidin, and delphinidin), flavanones (hesperetin, naringenin), and flavones (apigenin, luteolin).[6,18]

Multiple therapeutic events induced by these compounds include antioxidative activity, free radical removal, metal chelation, anticholinesterase activity, antiaging, neuroprotective, anti-inflammatory and neurotrophic roles.[6,15–17,19–37] Several lines of evidence suggested that TCM with copious flavonoids such as quercetin, apigenin, epigallocatechin-3-gallate, catechin, epigallocatechin, epicatechin-3-O-gallate, icariin, procyanidin, and silibinin act as acetylcholinesterase inhibitors, thus playing an essential role in AD therapeutics.[6,38]

Epigallocatechin-3-Gallate

Epigallocatechin-3-gallate (EGCG), an essential ingredient of green tea, has been extensively studied due to its remarkable antioxidant and anti-inflammatory properties.[39] It showed decreased AChE activity, glutathione peroxidase activity, and ROS.[4,40] Intraperitoneal injection of EGCG altered tau protein and ameliorated cognition in transgenic (Tg) mice.[4,41] Moreover, EGCG also curtailed cellular holo-amyloid precursor protein (APP) by potential iron chelation activity.[4] It was also found to ameliorate LPS-provoked apoptosis and memory dysfunction through nonamyloidogenic

Alzheimer's Disease Theranostics. https://doi.org/10.1016/B978-0-12-816412-9.00004-5
Copyright © 2019 Elsevier Inc. All rights reserved.

proteolysis by declining APP expression, beta-site APP cleaving enzyme 1 (BACE1) activity, and Aβ levels.[4] It also curtailed astrocyte activation and inflammatory factors such as tumor necrosis factor (TNF-α), interleukin-1 β (IL1-β), Interleukin-6 (IL-6).[4,42,43] Conversely, EGCG high-dose dietary supplementation showed initiation of inflammatory factors such as TNF-α, IL1-β, IL-6, and lipid inflammatory mediator PGE2 in mice.[4,44]

RESVERATROL

Resveratrol (3,5,4′-trihydroxy-trans-stilbene) (Resv), a polyphenolic compound abundantly found in grape skin and seeds, has plethora of therapeutic effects such as anti-inflammatory, antioxidative, and cardioprotective and neuroprotective functions.[45–48] Growing lines of evidence showed that Resv has robust ROS scavenging activity, iron chelating efficacy[49] and potential to curtail amyloid production.[50,51] Although Resv plays an essential role in ameliorating AD, the underlying mechanism is yet unknown.[51] It showed a substantial therapeutic role in AD treatment by regulating miRNAs and autophagy processes.[46] Low water solubility and short half-life limits its success although it shows significant therapeutic efficacy.[52,53] Despite better gastrointestinal absorption, it shows less bioavailability because of its speedy metabolism and removal from systemic circulation.[4] Resv augments intracellular calcium in cortical neurons by regulating secondary messengers, cGMP, cAMP, and NO, which also improve glucose consumption in cells.[4,54]

Resv reduced Aβ levels by provoking nonamyloidogenic pathway and enhancing Aβ scavenging.[4,27,28] More interestingly, binding specificity exhibited by Resv such as high affinity to fibrillary form in Aβ1-42 form and monomeric form in Aβ1-40 paved a way to streamline novel detection methods.[4,55] Numerous drug delivery systems employed to overcome the challenges associated with Resv delivery include lipid nano carriers,[56] solid lipid nanocarriers,[53] and chitosan nanoparticles.[57] Resv loaded solid lipid NPs showed potential protective efficacy against breast cancer in MDA-MB-231 cells.[53] Resv encapsulated lipid core nanocapsules profoundly enhanced brain targeting, thus remarkably enhancing therapeutic efficacy.[56,58]

SAIKOSAPONINS

Saikosaponins extracted from *Bupleuri Radix* showed enormous therapeutic potential against manifold age-related disease such as cancer and AD.[1] They exhibit antioxidant, anti-inflammatory, antibacterial, antiviral, and anticancer effects.[1,59–63] More particularly, saikosaponin C has been found to attenuate Aβ$_{1-40}$ and Aβ$_{1-42}$ production and tau phosphorylation.[1,64]

Bilobalide (BB), the vital terpenoid of *Ginkgo biloba*, attenuated the β secretase activity of cathepsin D and deteriorated the production of 2 β secretase cleavage compounds of APP, Aβ, and soluble APPβ, by influencing PI3K-dependent pathway.[4,65]

CURCUMIN

The rhizome of Curcuma longa generally known as turmeric contains a yellow pigment curcumin, a highly active agent of this herbal medicine. Curcumin, a vital ingredient of popular curry spice turmeric shows exemplary antioxidant, anti-inflammatory, antiangiogenic, and anticancer properties. Epidemiological studies showed that the prevalence of AD is significantly less in regions where it is used as a spice.[66] Several strategies based on its neuroprotective effects against neurodegenerative disorders include inhibition of enzymatic activity, antioxidative activity, anti-inflammatory activity, and reduction of cytotoxicity.

It plays a pivotal role in improving mitochondrial membrane potential in turn reducing ROS production and attenuating apoptotic cell death in Aβ25–35 treated neurons.[67] In addition, it also improves mitochondrial function and gene and protein expression in the murine brain with apolipoprotein E3 genotype carriers [2014]. Wang et al.[68] reported the substantial neuroprotective role of curcuminoids such as diarylalkyls curcumin (CCN), demethoxycurcumin (DMCCN), and bisdemethoxycurcumin (BDMCCN) in AD Drosophila melanogaster models via inhibition of β-amyloid precursor cleavage enzyme. Curcumin topped the list of 214 antioxidant compounds studied for preventing Aβ fibrils formation. The architecture of this curcumin plays a pivotal role in its ability to span BBB and bind to Aβ. The remarkable bioactivity of 1000 folds of a water-soluble curcumin conjugate (curcumin tethered to a sugar moiety) compared to curcumin emphasizes that a little modification to it can render this an appropriate therapeutic strategy for AD and other diseases.[69]

CLICK CHEMISTRY

Curcumin nanoliposomes designed by click chemistry method or conjugation with phospholipid succeeded in stability *in vitro* and *in vivo* in turn showing remarkably high and hitherto unidentified affinity to

Aβ1–42.[70] However, increased curcumin dosage shows adverse effects such as mitochondrial and nuclear DNA damages to human hepatoma G2 cells and remarkable ROS elevation. Despite the significant theranostic effect of curcumin, it has limitations for human use presently. McClure et al.[70] designed an innovative inhalable curcumin formulation which in 5XFAD mouse model binds to the amyloid plaques and substantially improves the symptoms of the disease. More recently Hucklenbroich et al.[71] discovered a novel robust aromatic tumerone which induced a neuroprotective role both *in vitro* and *in vivo* through the proliferation of neural stem cells. This novel compound tangibly becomes the cornerstone based on which substantial armamentarium for several neurodegenerative disorders can be streamlined. SLN-loaded curcumin showed potential recuperation of membrane lipids and AChE activity in $AlCl_3$-treated mice, and the results were on par with those obtained by rivastigmine.[4,72]

NANOTECHNOLOGY

Mounting evidence suggested that quercetin nanoparticle showed better oral absorption without perturbance in coordination or locomotor activity and enhanced brain loading.[6,73] Quercetin, a polyphenolic flavonoid, chiefly exerts protective efficacy by ROS elimination and antioxidant activity.[4,74] The nanoparticle (NP) loaded quercetin was found to ameliorate memory and learning significantly when evaluated by the Morris water maze test. Additionally, remarkable decrease in glial fibrillary acidic protein (GFAP) was found in hippocampus, which was not shown by quercetin alone.[6,73] In another study, it has been shown that quercetin NPs profoundly curtailed malonaldehyde levels and augmented levels of glutathione peroxidase and catalase in the brain, thus enhancing spatial memory in scopolamine-induced memory deficits in animals.[6,75] In another study, quercetin-loaded solid lipid nanoparticles (SLNs) showed enhanced brain spanning of the compound and significantly high therapeutic efficacy.[4,76]

Anthocyanin loaded poly(ethylene) glycol-AuNPs showed enhanced localization in brain and decreased β-amyloid and β-amyloid cleaving enzyme (BACE 1), tau hyperphosphorylation.[6,77]

CHITOSAN

Chitosan is a polysaccharide which comprises irregularly dispersed β-(1–4) linked glucosamine, and N-acetyl-D-glucosamine is a component of exoskeleton of crustaceans.[4] Several lines of evidence showed that chitosan

alone offers neuroprotective efficacy against H_2O_2 instigated apoptosis by averting Aβ formation and curbing intrinsic apoptosis pathway. With ample amenable protective properties chitosan has been extensively used as substantial drug delivery system (DDS).[4,78] Chitosan NPs were used to target rivastigmine and tacrine to the brain to overcome AD.[4,79–82]

CONCLUSION

Despite demonstration of remarkable therapeutic efficacy of AD drugs in preclinical trials, growing number of studies fail in clinical trials due to their low spanning ability through BBB.[4] Interestingly, some of the compounds emerged as substantial therapeutic molecules despite the failure of manifold compounds.[4,83,84]

Despite the significant role of galanthamine in AD treatment, its low availability limits its success.[14] Studies have proven that galanthamine obtained only from plants is insufficient to meet the current demand.[14,85] Therefore, it is essential to focus on multifarious genes and enzymes entailed in galanthamine biosynthesis.[14] Furthermore, there is a significant dearth in understanding the underlying mechanisms of action of natural compounds. Although saikosaponins showed ample therapeutic effects against various age-related diseases, their cytotoxicity in normal cells has not yet been well established. Therefore, cytotoxicity studies are warranted.

REFERENCES

1. Kim BM. The role of saikosaponins in therapeutic strategies for age-related diseases. *Oxid Med Cell Longev*. 2018;2018:8275256.
2. Anekonda TS, Reddy PH. Can herbs provide a new generation of drugs for treating Alzheimer's disease? *Brain Res Brain Res Rev*. 2005;50:361–376.
3. Farver D. The use of "natural products" in clinical medicine. *S D J Med*. 1996;49:129–130.
4. Ansari, N., Khodagoli, F. Natural products as promising drug candidates for the treatment of Alzheimer's disease: Molecular mechanism aspect. *Curr Neuropharmacol*. 11:414–429.
5. Balachandran P, Govindarajan R. Ayurvedic drug discovery. *Expet Opin Drug Discov*. 2007;2:1631–1652.
6. de Andrade Teles RB, Diniz TC, Costa Pinto TC, et al. Flavonoids as therapeutic agents in Alzheimer's and Parkinson's diseases: a systematic review of preclinical evidences. *Oxid Med Cell Longev*. 2018. [in press].
7. Wu WY, Hou JJ, Long HL, Yang WZ, Liang J, Guo DA. TCM-based new drug discovery and development in China. *Chin J Nat Med*. 2014;12:241–250.

8. Ren ZL, Zuo PP. Neural regeneration: role of traditional Chinese medicine in neurological diseases treatment. *J Pharm Sci.* 2012;120:139–145.

9. Wang Z, Wan H, Li J, Zhang H, Tian M. Molecular imaging in traditional Chinese medicine therapy for neurological diseases. *BioMed Res Int.* 2013;2013:608430.

10. Irwin RL, Smith HJ. Cholinesterase inhibition by galanthamine and lycoramine. *Biochem Pharmacol.* 1960;3:147–148.

11. Proskurnina NF, Yakovleva AP. About alkaloids of Galanthus worownii. 2. The isolation of a new alkaloid. *Zurnal Obshchei Khimii.* 1952;22:1899–1902.

12. Proskurnina NF, Yakovleva AP. About alkaloids of Galanthus worownii. 3. About galanthamine structure. *Zurnal Obshchei Khimii.* 1955;25:1035–1039.

13. Woodruff-Pak DS, Vogel RW, Wenk GL. Galantamine: effect on nicotinic receptor binding, acetylcholinesterase inhibition, and learning. *Proc Natl Acad Sci USA.* 2001:2089–2094.

14. Li W, Yang Y, Qiao C, et al. Functional characterization of phenylalanine ammonia-lyase- and cinnamate 4-hydroxylase-encoding genes from Lycoris radiata, a galanthamine-producing plant. *Int J Biol Macromol.* 117:1264–1279.

15. Grosso C, Valentao P, Ferreres F, Andrade P. The use of flavonoids in central nervous system disorders. *Curr Med Chem.* 2013;20(37):4694–4719.

16. Liu CM, Ma JQ, Liu SS, Zheng GH, Feng ZJ, Sun JM. Proanthocyanidins improves lead-induced cognitive impairments by blocking endoplasmic reticulum stress and nuclear factor-κB-mediated inflammatory pathways in rats. *Food Chem Toxicol.* 2014;72:295–302.

17. Nabavi SF, Braidy N, Habtemariam S, et al. Neuroprotective effects of chrysin: from chemistry to medicine. *Neurochem Int.* 2015;90:224–231.

18. Ferreyra MLF, Rius SP, Casati P. Flavonoids: biosynthesis, biological functions, and biotechnological applications. *Front Plant Sci.* 2012;3:222.

19. Adedayo BC, Oboh G, Oyeleye S, Ejakpovi II, Boligon AA, Athayde ML. Blanching alters the phenolic constituents and *in vitro* antioxidant and anticholinesterases properties of fireweed (Crassocephalum crepidioides). *J Taibah Univ Med Sci.* 2015;10:419–426.

20. Ahmed ME, Khan MM, Javed H, et al. Amelioration of cognitive impairment and neurodegeneration by catechin hydrate in rat model of streptozotocin-induced experimental dementia of Alzheimer's type. *Neurochem Int.* 2013;62:492–501.

21. Ashafaq M, Raza SS, Khan MM, et al. Catechin hydrate ameliorates redox imbalance and limits inflammatory response in focal cerebral ischemia. *Neurochem Res.* 2012;37(8):1747–1760.

22. Brown RC, Lockwood AH, Sonawane BR, et al. Neurodegenerative diseases: an overview of environmental risk factors. *Environ Health Perspect.* 2005;113:1250–1256.

23. Elbaz A, Carcaillon L, Kab S, Moisan F. Epidemiology of Parkinson's disease. *Rev Neurol.* 2016;172:14–26.

24. Kelsey NA, Wilkins HM, Linseman DA. Nutraceutical antioxidants as novel neuroprotective agents. *Molecules.* 2010;15:7792–7814.

25. Khan MM, Kempuraj D, Thangavel R, Zaheer A. Protection of MPTP-induced neuroinflammation and neurodegeneration by pycnogenol. *Neurochem Int.* 2013;62:379–388.

26. Kim DH, Kim S, Jeon SJ, et al. Tanshinone I enhances learning and memory, and ameliorates memory impairment in mice via the extracellular signal-regulated kinase signaling pathway. *Brit J Pharmacol.* 2009;158:1131–1142. 2009.

27. Li F, Gong Q, Dong H, Shi J. Resveratrol, a neuroprotective supplement for Alzheimer's disease. *Curr Pharm Des.* 2012;18:27–33.

28. Li JK, Jiang ZT, Li R, Tan J. Investigation of antioxidant activities and free radical scavenging of flavonoids in leaves of polygonum multiflorum thumb. *China Food Additives.* 2012;2:69–74.

29. Li Q, Zhao H, Zhao M, Zhang Z, Li Y. Chronic green tea catechins administration prevents oxidative stress-related brain aging in C57BL/6J mice. *Brain Res.* 2010;1353:28–35.

30. Lin L, Ni B, Lin H, et al. Traditional usages, botany, phytochemistry, pharmacology and toxicology of Polygonum multiflorum Thunb.: a review. *J Ethnopharmacol.* 2015;159:158–183.

31. Matsuzaki K, Miyazaki K, Sakai S, et al. Nobiletin, a citrus flavonoid with neurotrophic action, augments protein kinase A-mediated phosphorylation of the AMPA receptor subunit, GluR1, and the postsynaptic receptor response to glutamate in murine hippocampus. *Eur J Pharmacol.* 2008;578:194–200.

32. Nday CM, Halevas E, Jackson GE, Salifoglou A. Quercetin encapsulation in modified silica nanoparticles: potential use against Cu(II)-induced oxidative stress in neurodegeneration. *J Inorg Biochem.* 2015;145:51–64.

33. Prakash D, Sudhandiran G. Dietary flavonoid fisetin regulates aluminium chloride-induced neuronal apoptosis in cortex and hippocampus of mice brain. *J Nutr Biochem.* 2015;26:1527–1539.

34. Pringsheim T, Jette N, Frolkis A, Steeves TD. The prevalence of Parkinson's disease: a systematic review and meta-analysis. *Mov Disord.* 2014;29:1583–1590.

35. Spencer JP, Vafeiadou K, Williams RJ, Vauzour D. Neuroinflammation: modulation by flavonoids and mechanisms of action. *Mol Aspect Med.* 2012;33:83–97.

36. Tysnes OB, Storstein A. Epidemiology of Parkinson's disease. *J Neural Transm.* 2017;124:901–905.

37. Zhang Q, Zhao JJ, Xu J, Feng F, Qu W. Medicinal uses, phytochemistry and pharmacology of the genus Uncaria. *J Ethnopharmacol.* 2015;173:48–80.

38. Jiang Y, Gao H, Turdu G. Traditional Chinese medicinal herbs as potential AChE inhibitors for anti-Alzheimer's disease: a review. *Bioorg Chem.* 2017;75:50–61.

39. Hu B, Liu X, Zhang C, Zeng X. Food macromolecule based nanodelivery systems for enhancing the bioavailability of polyphenols. *J Food Drug Anal.* 2017;25:3–15.

40. Biasibetti R, Tramontina AC, Costa AP, et al. Green tea (-)epigallocatechin-3- gallate reverses oxidative stress and reduces acetylcholinesterase activity in a strepto-zotocin-induced model of dementia. *Behav Brain Res.* 2012;236C:186–193.

41. Rezai-Zadeh K, Arendash GW, Hou H, et al. Green tea epigallocatechin-3-gallate (EGCG) reduces beta-amyloid mediated cognitive impairment and modulates tau pathology in Alzheimer transgenic mice. *Brain Res.* 2008;1214:177–187.

42. Li R, Huang YG, Fang D, Le WD. (-)-Epigallocatechin gallate inhibits lipopolysaccharide-induced micro-glial activation and protects against inflammation-mediated dopaminergic neuronal injury. *J Neurosci Res.* 2004;78:723–731.

43. Lee YJ, Choi DY, Yun YP, Han SB, Oh KW, Hong JT. Ep-igallocatechin-3-gallate prevents systemic inflammation-induced memory deficiency and amyloidogenesis *via* its anti-neuroinflammatory properties. *J Nutr Biochem.* 2012;24:298–310.

44. Pae M, Ren Z, Meydani M, et al. Dietary supplementa-tion with high dose of epigallocatechin-3-gallate pro-motes inflammatory response in mice. *J Nutr Biochem.* 2012;23:526–531.

45. Baur JA, Sinclair DA. Therapeutic potential of resveratrol: the *in vivo* evidence. *Nat Rev Drug Discov.* 2006;5:493–506.

46. Kou X, Chen N. Resveratrol as a natural autophagy regula-tor for prevention and treatment of Alzheimer's disease. *Nutrients.* 2017;24:9.

47. Oomen CA, Farkas E, Roman V, van der Beek EM, Luiten PG, Meerlo P. Resveratrol preserves cerebrovascular den-sity and cognitive function in aging mice. *Front Aging Neu-rosci.* 2009;1:4.

48. Saiko P, Szakmary A, Jaeger W, Szekeres T. Resveratrol and its analogs: defense against cancer, coronary disease and neurodegenerative maladies or just a fad? *Mutat Res.* 2008;658:68–94.

49. Kitada M, Koya D. Renal protective effects of resveratrol. *Oxid Med Cell Longev.* 2013:568093.

50. Donmez G, Wang D, Cohen DE, Guarente L. SIRT1 suppresses beta-amyloid production by activating the alpha-secretase gene ADAM10. *Cell.* 2010;142:320–332.

51. Wang GH, Wang LH, Wang C, et al. Spore powder of Gan-oderma lucidum for the treatment of Alzheimer disease: A pilot study. *Medicine (Baltimore).* 2018;97:e0636.

52. Neves AR, Lucio M, Lima JL, Reis S. Resveratrol in me-dicinal chemistry: a critical review of its pharmacokinet-ics, drug-delivery, and membrane interactions. *Curr Med Chem.* 2012;19:1663–1681.

53. Wang W, Zhang L, Chen T, et al. Anticancer effects of res-veratrol-loaded solid lipid nanoparticles on human breast cancer cells. *Molecules.* 2017;22.

54. Zhang JQ, Wu PF, Long LH, et al. Resveratrol promotes cellular glucose utilization in primary cultured cortical neurons *via* calcium dependent signaling pathway. *J Nutr Biochem.* 2013;24:629–637.

55. Ge JF, Qiao JP, Qi CC, Wang CW, Zhou JN. The binding of resveratrol to monomer and fibril amyloid beta. *Neuro-chem Int.* 2012;61:1192–1201.

56. Montenegro L, Parenti C, Turnaturi R, Pasquinucci L. Res-veratrol-loaded lipid nanocarriers: correlation between *in vitro* occlusion factor and *in vivo* skin hydrating effect. *Pharmaceutics.* 2017;10:58.

57. Wu J, Wang Y, Yang H, Liu X, Lu Z. Preparation and bi-ological activity studies of resveratrol loaded ionically cross-linked chitosan-TPP nanoparticles. *Carbohydr Polym.* 2017;175:170–177.

58. Frozza RL, Bernardi A, Paese K, et al. Characterization of trans-resveratrol-loaded lipid-core nanocapsules and tissue distribution studies in rats. *J Biomed Nanotechnol.* 2010;6:694–703.

59. Sui C, Zhang J, Wei J, et al. Transcriptome analysis of Bu-pleurum chinense focusing on genes involved in the bio-synthesis of saikosaponins. *BMC Genomics.* 2011;12:539.

60. Wu GC, Wu L, Fan Y, Pan HF. Saikosaponins: a potential treatment option for systemic lupus erythematosus. *Irish J Med Sci.* 2011;180:259–261.

61. Wu SJ, Lin YH, Chu CC, Tsai YH, Chao JCJ. Curcumin or saikosaponin improves hepatic antioxidant capacity and protects against CCl$_4$-induced liver injury in rats. *J Med Food.* 2008;11:224–229.

62. Wu SJ, Tam KW, Tsai YH, Chang CC, Chao JCJ. Curcumin and saikosaponin a inhibit chemical-induced liver inflam-mation and fibrosis in rats. *Am J Chin Med.* 2010;38:99–111.

63. Zhang BZ, Guo XT, Chen JW, et al. Saikosaponin-D at-tenuates heat stress-induced oxidative damage in LLC-PK$_1$ cells by increasing the expression of anti-oxidant enzymes and HSP72. *Am J Chin Med.* 2014;42:1261–1277.

64. Lee TH, Park S, You MH, Lim JH, Min SH, Kim BM. A potential therapeutic effect of Saikosaponin C as a novel dual-target anti-Alzheimer agent. *J Neurochem.* 2016;136:1232–1245.

65. Shi C, Wu F, Xu J, Zou J, et al. Bilobalide regulates solu-ble amyloid precursor protein release *via* phosphatidyl inositol 3 kinase dependent pathway. *Neurochem Int.* 2011;59:59–64.

66. Sezgin Z, Dincer Y. Alzheimer's disease and epigenetic diet. *Neurochem Int.* 2014;78:105–116.

67. Sun, et al. Activation of SIRT1 by curcumin blocks the neurotoxicity of amyloid-β25-35 in rat cortical neurons. *Biochem Biophys Res Commun.* 2014;448:89–94.

68. Wang X, Kim JR, Lee SB, et al. Effects of curcuminoids iden-tified in rhizomes of Curcuma longa on BACE-1 inhibitory and behavioral activity and lifespan of Alzheimer's disease Drosophila models. *BMC Compl Altern Med.* 2014;14:88.

69. Lee WH, Loo CY, Bebawy M, et al. Curcumin and its de-rivatives: their application in neuropharmacology and neuroscience in the 21st century. *Curr Neuropharmacol.* 2013;11:338–378.

70. McClure R, Yanagisawa D, Stec D, et al. Inhalable cur-cumin: offering the potential for translation to imaging and treatment of Alzheimer's disease. *J Alzheimers Dis.* 2014;44:283–295.

71. Hucklenbroich, Klein R, Neumaier B, et al. Aromatic-turmerone induces neural stem cell proliferation *in vitro* and *in vivo*. *Stem Cell Res Ther*. 2014;5:100.

72. Alam S, Panda JJ, Chauhan VS. Novel dipeptide nanoparticles for effective curcumin delivery. *Int J Nanomed*. 2012;7:4207–4221.

73. Moreno LCGEI, Puerta E, Suarez-Santiago JE, Santos-Magalhaes NS, Ramirez MJ, Irache JM. Effect of the oral administration of nanoencapsulated quercetin on a mouse model of Alzheimer's disease. *Int J Pharm*. 2017;517:50–57.

74. Ossola B, Kaariainen TM, Mannisto PT. The multiple faces of quercetin in neuroprotection. *Expert Opin Drug Saf*. 2009;8:397–409.

75. Palle S, Neerati P. Quercetin nanoparticles attenuates scopolamine induced spatial memory deficits and pathological damages in rats. *Bull Fac Pharm Cairo Univ*. 2017;55:101–106.

76. Dhawan S, Kapil R, Singh B. Formulation development and systematic optimization of solid lipid nanoparticles of quercetin for improved brain delivery. *J Pharm Pharmacol*. 2011;63:342.

77. Kim MJ, Rehman SU, Amin FU, Kim MO. Enhanced neuroprotection of anthocyanin-loaded PEGgold nanoparticles against Aβ1-42-induced neuroinflammation and neurodegeneration via the NF-KB/JNK/GSK3β signaling pathway. *Nanomed Nanotechnol Biol Med*. 2017;13:2533–2544.

78. Mengatto LN, Helbling IM, Luna JA. Recent advances in chitosan films for controlled release of drugs. *Recent Pat Drug Deliv Formul*. 2012;6:156–170.

79. Fazil M, Md S, Haque S, et al. Development and evaluation of rivastigmine loaded chitosan nanoparticles for brain targeting. *Eur J Pharm Sci*. 2012;47:6–15.

80. Wilson B, Samanta MK, Muthu MS, Vinothapooshan G. Design and evaluation of chitosan nanoparticles as novel drug carrier for the delivery of rivastigmine to treat Alzheimer's disease. *Ther Deliv*. 2011;2:599–609.

81. Wilson B, Samanta MK, Santhi K, Kumar KP, Ramasamy M, Suresh B. Chitosan nanoparticles as a new delivery system for the anti-Alzheimer drug tacrine. *Nanomedicine*. 2010;6:144–152.

82. Wilson B, Samanta MK, Santhi K, Sampath Kumar KP, Ramasamy M, Suresh B. Significant delivery of tacrine into the brain using magnetic chitosan microparticles for treating Alzheimer's disease. *J Neurosci Methods*. 2009;177:427–433.

83. Frakey LL, Salloway S, Buelow M, Malloy P. A randomized, double-blind, placebo-controlled trial of modafinil for the treatment of apathy in individuals with mild-to-moderate Alzheimer's disease. *J Clin Psychiatry*. 2012;73:796–801.

84. Winblad B, Andreasen N, Minthon L, Floesser A, Imbert G, Dumortier T. Safety, tolerability, and antibody response of active A immunotherapy with CAD106 in patients with Alzheimer's disease: randomised, double-blind, placebo-controlled, first-in-human study. *Lancet Neurol*. 2012;11:597–604.

85. Czollner L, Frantsits W, Kuenburg B, Hedenig U, Frohlich J, Jordis U. New kilogram-synthesis of the anti-Alzheimer drug (–)-galanthamine. *Tetrahedron Lett*. 1998;39:2087–2088.

CHAPTER 5

Blood-Brain Barrier–Targeted Nanotechnological Advances

ABSTRACT

The advent of manifold therapeutic arsenals showed appreciable improvement in the therapeutics for several diseases. However, a few barriers limit the success of bionanomedicine significantly. Blood-brain barrier (BBB) plays a pivotal role in brain protection. However, when impaired, it leads to manifold neurodegenerative disorders such as Alzheimer's disease (AD), Parkinson's disease (PD), etc. Brain drug targeting has been a major challenge for a few decades despite the voluminous studies conducted until now. Panoply of strategies employed to achieve BBB permeability include but are not limited to the development of multifarious biomaterials using nanotechnology, surface functionalization with variety of targeting moieties, etc. We hereby focus on the status of BBB in physiology, AD pathology, and a few currently available nanotechnological advances to circumvent the challenges in drug delivery.

KEYWORDS

Alzheimer's disease; Blood-brain barrier; Matrix metalloproteinases; Nanotechnology; Surface functionalization.

INTRODUCTION

Edwin Goldman, an African-German, for the first time proved the presence of the blood-brain barrier (BBB) in 1909. In his study, a dye introduced into systemic circulation was spread all over the body except the brain and spinal cord, and the same when injected into the brain stained only the brain but not the other regions, implying that the existence of BBB limits the passage of dye.[1–3] The two major barriers of the brain are BBB and blood-cerebrospinal fluid barrier (BCSFB). BBB is the barrier between brain tissue and blood circulation, whereas BCSFB is the barrier between blood and cerebrospinal fluid (CSF).

BBB dysfunction has long been implicated in neurodegenerative disorders such as Alzheimer's disease (AD).[4] BBB impairment aids the entry of inflammatory mediators and immune cells into the brain.[4] A few possible ways to achieve brain entry are passive diffusion, receptor-mediated transcytosis, and carrier-mediated transcytosis.[5] While the conventional methods showed limited success, nanotechnological methods were explored considerably to achieve enhanced brain delivery of therapeutic molecules.

PHYSIOLOGY

The central components of BBB are the brain capillary endothelial cells (ECs), pericytes that feed ECs, astrocytes, microglial cells, and neurons.[3] BBB is highly enriched with mitochondrial content and devoid of fenestrations.[3,6] Various enzymes that constitute BBB include cholinesterases, γ-aminobutyric acid transaminases, aminopeptidases, and endopeptidases.[7]

BBB plays an essential role in the central nervous system (CNS) homeostasis by mediating ion balance and nutritional transport and averting entry of toxic molecules from circulation.[8,9] Organs of the body have capillaries which comprise endothelial cells with pores to facilitate the movement of small molecules from blood to the organs. However, the brain is devoid of such pores and instead contains tight junctions (TJs) made up of several transmembrane proteins which restrict the paracellular pathway.[3,10,11] Unlike the permeable vessels of other peripheral organs, vessels of BBB impede the transport of polar molecules into the brain. However, the brain transportation can be accomplished by particular receptors present on the epithelium in physiological or pathological circumstances.[3,12,13]

Small molecules of <400 Da can trespass BBB.[7,14,15] However, large molecules need specific receptors to pass through BBB. The primary transport mechanisms of BBB are peptide-mediated transport mechanisms

Alzheimer's Disease Theranostics. https://doi.org/10.1016/B978-0-12-816412-9.00005-7
Copyright © 2019 Elsevier Inc. All rights reserved.

such as receptor-based, absorptive-based, carrier-based, and nonspecific passive diffusion, and endocytosis.[3,16] A few key transporters at the BBB include glucose transporters and ATP binding cassette transporters such as P-glycoprotein[17,18]. TJs being complex mixtures of proteins such as junctional adhesion molecules (JAMs), occludin, claudin, and membrane-linked guanylate kinase exert transendothelial electrical resistance of 1500–2000 cm^2 and prevents the entry of ions and solutes.[3,19,20]

PATHOLOGY

Higher oxygen demand, considerably low antioxidant activity, and presence of more lipid cells make the brain vulnerable to the ROS-induced oxidative stress.[21–23] Additionally, etiological factors that provoke BBB dysfunction are immune cells, pathogens, and drugs.[3] BBB has long been considered to be the primary hurdle in the treatment of brain disorders such as AD.[5,24,25] Its impairment has been considered the earliest event in AD even before Aβ aggregation and neurofibrillary tangle (NFT) formation.[26] Impaired cerebral blood flow enhances Aβ aggregation, influx of antibodies and neuroinflammation, eventually leading to AD.[27,28] Augmented BBB permeability leads to AD and related dementias although unrelated to amyloid pathology.[9] Studies showed an intact BBB in murine models with noteworthy AD pathology corroborating these findings.[9,29] In addition, a few preclinical and human disease studies showed apolipoprotein's (ApoE4) role in BBB dysfunction. For example, studies on transgenic mice exhibited ApoE4 generation by astrocytes which resulted in BBB impairment.[9,30] Conversely, a few studies showed no link between ApoE and BBB dysfunction.[9,29,31,32]

BBB permeability assessment methods primarily include analysis of the CSF/blood albumin ratio, histologic identification of the proteins obtained from blood in the brain tissue, and brain imaging such as magnetic resonance imaging (MRI) or positron emission tomography (PET).[9] BBB dysfunction also provokes oxidative stress and chronic inflammation in endothelial cells.[9,33,34] Aβ aggregation in the vascular wall probably damages endothelial cells, thus causing BBB dysfunction and AD.[9,35,36] Moreover, altered vascular endothelial growth factor (VEGF) production plays a probable role in BBB dysfunction.[9,37]

CSF analysis accentuated that BBB impairment in AD takes place via injury to the pericytes.[3,38] BBB impairment provokes microglial activation and reactive astrogliosis, which in turn augment expression and/or release of high mobility group box protein 1 and thrombin.[26] *PICALM*, the gene encoding phosphatidylinositol binding clathrin assembly (PICALM), has been found to curtail Aβ scavenging in the murine BBB and enhanced Aβ pathology.[39] It is profusely expressed on capillary endothelium of BBB and facilitates Aβ scavenging from the brain.[39–41,41a] Based on these findings, it can be concluded that PICALM is a substantial therapeutic target in the treatment of AD. Nevertheless, the molecular underpinnings entailed in these events are yet to be elucidated.[26]

BCSFB

L-Glutamate plays a key role in cognition. Therefore, the presence of *N*-Methyl-D-aspartate glutamate receptor (NMDA-R) and deteriorated BCSFB lead to dementia such as AD.[27,42] Its impairment also occurs via acute infections or revival of mycoplasma, herpes simplex virus (HSV), and Epstein-Bar virus (EBV).[27,43–45] Therefore, these infections may play a central role in inducing AD.[27]

THERAPEUTICS

Several lines of evidence suggested that approximately 98% of the drugs cannot pass through BBB.[7,14] Numerous antioxidant compounds were discovered to ameliorate AD symptoms. For example, protocatechuic acid (PCA), being the vital component among the polyphenolic compounds, exhibits longer systemic circulation, better BBB permeability, and robust therapeutic efficacy against AD.[23] Recent *in vitro* studies showed the robust therapeutic efficacy of PCA in attenuating glutamate release, reactive oxygen species (ROS) production, caspase-3 activation, and curtailing Aβ-induced apoptosis by perturbing β amyloid fibrils.[46–48]

Lipidization, a process of making drugs more lipophilic, also ameliorates BBB perviousness. For example, to make morphine BBB permeable, it is converted to heroin by acetylation of its two hydroxyl groups.[7,49] The disadvantage of this process is enhanced permeability of nonspecific drugs in other organs as well.[7,15] Although vasoactive molecules such as leukotriene C4 and bradykinin are employed to interrupt BBB, they appear to provoke complete damage of brain tissue.[7,50,51]

Matrix Metalloproteinases

Matrix metalloproteinases (MMPs) are the proteases which degenerate extracellular matrix and TJs of endothelial cells in AD.[52] They were found to attach to occludin and impair TJs, thus enhancing the BBB permeability in AD.[3,53,54] Accumulating evidence showed remarkable reduction in claudin 1 and claudin 5 and appreciable augmentation in MMP2 and MMP9 in rat brain microvessels after treatment with higher concentrations of Aβ 40.[52,55]

Nanoparticles

Because transport of large size molecules within the extracellular space of the brain is a major biological riddle, ideally NPs below 100 nm diameter can accomplish enhanced brain targeting.[56,57] In contrast, small particles within 10 nm may show less binding affinity and may be expelled from the systemic circulation even before getting absorbed.[57–59] Plethora of *in vitro* and *in vivo* studies showed the influence of NP size in spanning BBB and biodistribution in mice.[57,60–62] BBB passage can be accomplished by potential inhibition of efflux system by conjugating polysorbate 80 on NPs.[63–65a] Array of brain drug delivery devices include

DDS exploiting NPs, liposomes, and dendrimers[5,66,67] (Table 5.1). NPs can ferry through BBB via the TJs of endothelial cells.[65a,68,69]

Brambilla et al. developed poly[(hexadecyl cyanoacrylate)-co-methoxypoly(ethylene glycol) cyanoacrylate] NPs which showed deterioration of Aβ aggregation *in vitro*.[65a,70,71] Recently designed ginsenoside Rg3 and Thioflavin T loaded poly(lactic-co-glycolic acid) (PLGA) nanoparticles (NPs) conferred significant efficacy in BBB entry and reduction of Aβ in *in vitro* model.[72,73] designed multifunctional NPs using PEGylated dendrigraft poly-L-lysines (DGLs) which when systemically administered showed substantial codelivery of therapeutic gene and peptide. While the noncoding RNA plasmid attenuated the activity of vital enzyme in Aβ formation, the therapeutic D-peptide enables deterioration of NFTs.[73] The numerous amine groups of DGLs exhibit copious modification sites and positive charge in physiological pH[73] (*in vivo*).

TABLE 5.1
Multifarious Brain Delivery Systems Targeting Blood-Brain Barrier (BBB)

Model	Strategy	Outcome	References
In vitro	FITC-loaded ferritins	Enhanced BBB permeation of ferritins	91
	Poly[(hexadecyl cyanoacrylate)-co-methoxypoly(ethylene glycol) cyanoacrylate]	Deterioration of Aβ fibrillation	70,71
	Ginsenoside Rg3 and thioflavin T loaded poly(lactic-co-glycolic acid) (PLGA) NPs	Reduction of Aβ levels	72
	Andrographolide-loaded human serum albumin NPs (HSAT NPs)	Enhanced bioavailability and BBB ferrying	82
In vivo	PEGylated dendrigraft poly-L-lysines (DGLs)	Deterioration of NFTs	73
	Poly(n-butyl cyanoacrylate) (PBCA) NPs coated with polysorbate 80	Targeted delivery of rivastigmine	74
	Apo E3 and Aβ targeted antibodies used as ligands	Enhance BBB delivery	81
	Insulin-conjugated gold NPs (INS-GNPs)	Noninvasive identification of particle accumulation	57
	Glutathione functionalized PEGylated liposomes	Enhanced brain biodistribution	87
MNPs			
	Magnetic nanocontainers	Effective BBB passage	88
	PEG and monodispersed nitrodopamine conjugated MNPs	AβO binding	89
	Osmotin-loaded MNPs	Amelioration of synaptic defects and Aβ aggregation	90

Wilson et al.[74] synthesized poly(n-butyl cyanoacrylate) (PBCA) NPs coated with polysorbate 80 to achieve targeted delivery of rivastigmine and treat AD in mice.[65a] The proposed mechanism for NP's entry in this study was potential interaction of polysorbate 80 with endothelial cells of BBB. Sun et al.[75] studied the substantial role of polysorbate 80 in initiating BBB entry of the molecules (*in vivo*).[65a] In another study, idebenone-loaded chitosan and N-carboxymethyl chitosan NPs were designed using the spray drying method. Despite no testing of these NPs against AD, overarching evidence exhibited the antioxidant potential of idebenone in the treatment of AD.[65a,76–78] Apo E3 and antibodies targeted at Aβ are extensively employed ligands to overcome BBB impediment.[79–81]

Andrographolide, a substantial therapeutic diterpenoid abundantly found in the Asian medicinal plant Andrographis paniculata, exhibits low bioavailability. With a view to overcoming this issue, Guccione et al.[82] developed andrographolide-loaded human serum albumin NPs (HSAT NPs) and poly ethyl cyanoacrylate NPs (PECA NPs) and evaluated their BBB ferrying efficacy. According to their studies, HSAT NPs enhanced BBB perviousness twofold compared to free andrographolide, whereas PECA NPs induced only a little perturbance in BBB.[82] Despite overarching evidence for the efficacy of multifarious NPs in overcoming AD, a few studies showed negative impact on the BBB and resulted in AD induction. For example, silver NPs (AgNPs) were found to accumulate in TJs of N2a cells after impairment of proteins such as claudin-5, ZO-1 and promoted the expression of amyloid precursor protein (APP) eventually leading to neuronal apoptosis.[83]

Gold nanoparticles

Betzer et al.[57] designed insulin conjugated gold NPs (INS-GNPs) and accomplished targeting to the particular brain regions. According to their study, INS-GNPs with 20 nm successfully spanned BBB via insulin receptors on BBB and acted as computed tomography contrast agents, thus facilitating noninvasive identification of particle accumulation [57,84] (*In Vivo*). GNPs employed in this study render remarkable features such as size compliance, enhanced circulation time, stability, and biosafety.[57,85,86] In line with this, it can be understood that the GNP can form a robust core which can facilitate multifarious ligand conjugation and accomplish manifold therapeutic effects.[57]

Liposomes

Liposomes are considered amenable DDS in brain targeting since their phospholipid bilayer resembles the physiological cell membrane.[5] With a view to accomplishing pronounced brain targeting, surface functionalization of liposomes with ligands such as lactoferrin, transferrin, glutathione, and glucose was done and used to treat AD.[5] Rip et al.[87] designed a glutathione functionalized PEGylated liposomes and studied their brain distribution using carboxyfluorescein, a fluorescent particle (*in vivo*).

Magnetic nanoparticles

Magnetic nanocontainers with electromagnetic fields of 28 mT (0.43 T/m) and 79.8 mT (1.39 T/m) were developed and injected via tail vein which showed effective BBB passage, thus ameliorating AD symptoms in mice.[88] In another study, PEG and monodispersed nitrodopamine conjugated MNPs were developed which can alter their surface via carboxylation and can further conjugate to AβO-specific antibodies.[89] More recently, osmotin-loaded magnetic NPs were designed which exhibited enhanced BBB crossing and amelioration of synaptic defects and Aβ aggregation.[90] In this study, fluorescent carboxyl magnetic Nile Red particles (FMNPs) were designed which were driven even to hippocampus and cortex regions in mice via functionalized magnetic field (FMF).[90]

CONCLUSIONS AND FUTURE PERSPECTIVES

BBB plays a pivotal role in normal physiology of brain and renders remarkable protection by preventing the entry of pathogens. However, it has also been considered a major barrier and a biological riddle in the treatment of neurodegenerative disorders such as AD. Wealth of studies targeting conventional treatment options such as antioxidant, anti-inflammatory, and cholinesterase inhibition therapies against AD yielded limited success. To circumvent these issues, manifold nanotechnological methods have been extensively employed. Despite overarching studies and their efficacy in enhancement of drug delivery through BBB, their clinical translation has been a herculean task till date. Moreover, a few strategies employed to accomplish BBB ferrying may show impending adverse effects such as permeation of toxic components into the brain, etc. Therefore, substantial future nanotechnological studies are warranted to accomplish the effective BBB and AD therapy.

REFERENCES

1. Alavijeh MS, Chishty M, Qaiser MZ, Palmer AM. Drug metabolism and pharmacokinetics, the blood-brain barrier, and central nervous system drug discovery. *NeuroRx.* 2005;2:554–571.

2. Clarke E, O'Malley CD. *The Human Brain and Spinal Cord. A Historical Study Illustrated by Writings from Antiquity to the Twentieth Century.* Berkeley: University of California Press; 1968.

3. Erdo F, Denes L, de Lange E. Age-associated physiological and pathological changes at the blood-brain barrier: a review. *J Cereb Blood Flow Metab.* 2017;37:4–24.

4. Le Page A, Lamoureux J, Bourgade K, et al. Polymorphonuclear neutrophil functions are differentially altered in amnestic mild cognitive impairment and mild Alzheimer's disease patients. *J Alzheimers Dis.* 2017;60:23–40.

5. Agrawal M, Ajazuddin, Tripathi DK, et al. Recent advancements in liposomes targeting strategies to cross blood-brain barrier (BBB) for the treatment of Alzheimer's disease. *J Contr Rel.* 2017;260:61–77.

6. Sanchez-Covarrubias L, Slosky LM, Thompson BJ, et al. Transporters at CNS barrier sites: obstacles or opportunities for drug delivery? *Curr Pharm Des.* 2014;20:1422–1449.

7. Dube T, Chibh S, Mishra J, Panda JJ. Receptor targeted polymeric nanostructures capable of navigating across the blood-brain barrier for effective delivery of neural therapeutics. *ACS Chem Neurosci.* 2017;8:2105–2117.

8. Chow BW, Gu C. The molecular constituents of the blood-brain barrier. *Trends Neurosci.* 2015;38:598–608.

9. Janelidze S, Hertze J, Nagga K, et al. Increased blood-brain barrier permeability is associated with dementia and diabetes but not amyloid pathology or APOE genotype. *Neurobiol Aging.* 2017;51:104–112.

10. Abbott NJ. Dynamics of CNS barriers: evolution, differentiation, and modulation. *Cell Mol Neurobiol.* 2005;25:5–23.

11. Ballabh P, Braun A, Nedergaard M. The blood-brain barrier: an overview: structure, regulation, and clinical implications. *Neurobiol Dis.* 2004;16:1–13.

12. Zlokovic BV, Apuzzo ML. Cellular and molecular neurosurgery: pathways from concept to reality–part I: target disorders and concept approaches to gene therapy of the central nervous system. *Neurosurgery.* 1997;40:789–803. discussion 803–804.

13. Zlokovic BV, Hyman S, McComb JG, et al. Kinetics of arginine-vasopressin uptake at the blood-brain barrier. *Biochim Biophys Acta.* 1990;1025:191–198.

14. Pardridge WM. The blood brain barrier: bottleneck in brain drug development. *NeuroRx.* 2005;2:3–14.

15. Pardridge WM. Blood-brain barrier delivery. *Drug Discov Today.* 2007;12:54–61.

16. Zlokovic BV. Cerebrovascular permeability to peptides: manipulations of transport systems at the blood-brain barrier. *Pharm Res.* 1995;12:1395–1406.

17. de Wit NM, Vanmol J, Kamermans A, Hendriks J, de Vries HE. Inflammation at the blood-brain barrier: the role of liver X receptors. *Neurobiol Dis.* 2017;107:57–65.

18. Persidsky Y, Ramirez SH, Haorah J, Kanmogne GD. Blood-brain barrier: structural components and function under physiologic and pathologic conditions. *J Neuroimmune Pharmacol.* 2006;1:223–236.

19. Hawkins BT, Davis TP. The blood-brain barrier/neurovascular unit in health and disease. *Pharmacol Rev.* 2005;57:173–185.

20. Zlokovic BV. The blood-brain barrier in health and chronic neurodegenerative disorders. *Neuron.* 2008;57:178–201.

21. Miller E, Morel A, Saso L, Saluk J. Isoprostanes and neuroprostanes as biomarkers of oxidative stress in neurodegenerative diseases. *Oxid Med Cell Longev.* 2014:1–10.

22. Kim GH, Kim JE, Rhie SJ, Yoon S. The role of oxidative stress in neurodegenerative diseases. *Exp Neurobiol.* 2015;24:325–340.

23. Krzysztoforska K, Mirowska-Guzel D, Widy-Tyszkiewicz E. Pharmacological effects of protocatechuic acid and its therapeutic potential in neurodegenerative diseases: review on the basis of *in vitro* and *in vivo* studies in rodents and humans. *Nutr Neurosci.* 2017:1–11.

24. Meeuwsen EJ, Melis RJ, Van Der Aa GC, et al. Effectiveness of dementia follow-up care by memory clinics or general practitioners: randomised controlled trial. *BMJ.* 2012;344:e3086.

25. Maussang D, Rip J, van Kregten J, et al. Glutathione conjugation dose-dependently increases brain-specific liposomal drug delivery *in vitro* and *in vivo. Drug Discov Today Technol.* 2016;20:59–69.

26. Festoff BW, Sajja RK, van Dreden P, Cucullo L. HMGB1 and thrombin mediate the blood-brain barrier dysfunction acting as biomarkers of neuroinflammation and progression to neurodegeneration in Alzheimer's disease. *J Neuroinflammation.* 2016;13:194.

27. Busse M, Kunschmann R, Dobrowolny H, et al. Dysfunction of the blood-cerebrospinal fluid-barrier and N-methyl-D-aspartate glutamate receptor antibodies in dementias. *Eur Arch Psychiatry Clin Neurosci.* 2017;268:483–492.

28. Kowal C, DeGiorgio LA, Nakaoka T, Hetherington H, Huerta PT, Diamond B, et al. Cognition and immunity; antibody impairs memory. *Immunity.* 2004;21:179–188.

29. Bien-Ly N, Boswell CA, Jeet S, et al. Lackof widespread BBB disruption in Alzheimer's disease models: focus on therapeuticantibodies. *Neuron.* 2015;88:289–297.

30. Bell RD, Winkler EA, Singh I, et al. Apolipoprotein E controls cerebrovascular integrity via cyclophilin A. *Nature.* 2012;485:512–516.

31. Bowman GL, Kaye JA, Moore M, Waichunas D, Carlson NE, Quinn JF. Blood-brain barrier impairment in Alzheimer disease: stability and functional significance. *Neurology.* 2007;68:1809–1814.

32. Karch A, Manthey H, Ponto C, et al. Investigating the association of ApoE genotypes with blood-brain barrier dysfunction measured by cerebrospinal fluid-serum albumin ratio in a cohort of patients with different types of dementia. *PLoS One.* 2013;8:e84405.

33. Di Marco LY, Venneri A, Farkas E, Evans PC, Marzo A, Frangi AF. Vascular dysfunction in the pathogenesis of Alzheimer's disease - a review of endothelium-mediated mechanisms and ensuing vicious circles. *Neurobiol Dis.* 2015;82:593–606.

34. Raz L, Knoefel J, Bhaskar K. The neuropathology and cerebrovascular mechanisms of dementia. *J Cereb Blood Flow Metab.* 2015;36:172–186.

35. Burgmans S, van de Haar HJ, Verhey FR, Backes WH. Amyloid-beta interacts with blood-brain barrier function in dementia: a systematic review. *J Alzheimers Dis.* 2013;35:859–873.

36. Erickson MA, Banks WA. Blood-brain barrier dysfunction as a cause and consequence of Alzheimer's disease. *J Cereb Blood Flow Metab.* 2013;33:1500–1513.

37. Reeson P, Tennant KA, Gerrow K, et al. Delayed inhibition of VEGF signaling after stroke attenuates blood-brain barrier breakdown and improves functional recovery in a comorbidity-dependent manner. *J Neurosci.* 2015;35:5128–5143.

38. Montagne A, Barnes SR, Sweeney MD, et al. Blood brain barrier breakdown in the aging human hippocampus. *Neuron.* 2015;85:296–302.

39. Zhao Z, Sagare AP, Ma Q, et al. Central role for PICALM in amyloid-β blood-brain barrier transcytosis and clearance. *Nat Neurosci.* 2015;18:978–987.

40. Deane R, et al. LRP/amyloid beta–peptide interaction mediates differential brain efflux of Abeta isoforms. *Neuron.* 2004;43:333–344.

41. Shibata M, Yamada S, Kumar SR, et al. Clearance of Alzheimer's amyloid-ss(1–40) peptide from brain by LDL receptor–related protein-1 at the blood–brain barrier. *J Clin Invest.* 2000;106:1489–1499.

41a. Zlokovic BV. Neurovascular pathways to neurodegeneration in Alzheimer's disease and other disorders. *Nat. Rev. Neurosci.* 2011;12:723–738.

42. Lau CG, Zukin RS. NMDA receptor trafficking in synaptic plasticity and neuropsychiatric disorders. *Nature Rev.* 2007;8:413–426.

43. Gable MS, Gavali S, Radner A, et al. Anti-NMDA receptor encephalitis: report of ten cases and comparison with viral encephalitis. *Eur J Clin Microbiol Infect Dis.* 2009;28:1421–1429.

44. Xu CL, Liu L, Zhao WQ, et al. Anti-*N*-methyl-d-aspartate receptor encephalitis with serum anti-thyroid antibodies and IgM antibodies against Epstein-Barr virus viral capsid antigen: a case report and one year follow-up. *BMC Neurol.* 2011;11:149.

45. Hacohen Y, Deiva K, Pettingill P, et al. *N*-Methyl-d-aspartate receptor antibodies in post-herpes simplex virus encephalitis neurological relapse. *Mov Disord.* 2013;29:90–96.

46. Yin X, Zhang X, Lv C, et al. Protocatechuic acid ameliorates neurocognitive functions impairment induced by chronic intermittent hypoxia. *Sci Rep.* 2015;5:14507.

47. Ban JY, Cho SO, Jeon SY, Bae K, Song KS, Seong YH. 3,4-Dihydroxybenzoic acid from Smilacis chinae rhizome protects amyloid β protein (25–35)-induced neurotoxicity in cultured rat cortical neurons. *Neurosci Lett.* 2007;420:184–188.

48. Hornedo-Ortega R, Alvarez-Fernandez MA, Cerezo AB, Richard T, Troncoso AM, Garcia Parrilla MC. Protocatechuic acid: inhibition of fibril formation, destabilization of preformed fibrils of amyloid-β and α-synuclein, and neuroprotection. *J Agric Food Chem.* 2016;64:7722–7732.

49. Oldendorf WH, Hyman S, Braun L, Oldendorf SZ. Blood-brain barrier: penetration of morphine, codeine, heroin, and methadone after carotid injection. *Science.* 1972;178:984–986.

50. Hashizume K, Black KL. Increased endothelial vesicular transport correlates with increased blood-tumor barrier permeability induced by bradykinin and leukotriene C4. *J Neuropathol Exp Neurol.* 2002;61:725–735.

51. Borlongan CV, Emerich DF. Facilitation of drug entry into the CNS via transient permeation of blood-brain barrier: laboratory and preliminary clinical evidence from bradykinin receptor agonist. *Cereport Brain Res Bull.* 2003;60:297–306.

52. Weekman EM, Wilcock DM. Matrix metalloproteinase in blood-brain barrier breakdown in dementia. *J Alzheimers Dis.* 2016;49:893–903.

53. Rosenberg GA, Yang Y. Vasogenic edema due to tight junction disruption by matrix metalloproteinases in cerebral ischemia. *Neurosurg Focus.* 2007;22:E4.

54. Yang Y, Rosenberg GA. MMP-mediated disruption of claudin-5 in the blood-brain barrier of rat brain after cerebral ischemia. *Methods Mol Biol.* 2011;762:333–345.

55. Hartz AM, Bauer B, Soldner EL, et al. Amyloid-beta contributes to blood-brain barrier leakage in transgenic human amyloid precursor protein mice and in humans with cerebral amyloid angiopathy. *Stroke.* 2012;43:514–523.

56. Patel T, Zhou J, Piepmeier JM, Saltzman WM. Polymeric nanoparticles for drug delivery to the central nervous system. *Adv Drug Deliv Rev.* 2012;64:701–705.

57. Betzer O, Shilo M, Opochinsky R, et al. The effect of nanoparticle size on the ability to cross the blood-brain barrier: an *in vivo* study. *Nanomedicine.* 2017;12:1533–1546.

58. Gao H, Shi W, Freund LB. Mechanics of receptor-mediated endocytosis. *Proc Natl Acad Sci USA.* 2005;102:9469–9474.

59. Yoo J. Factors that control the circulation time of nanoparticles in blood: challenges, solutions and future prospects. *Curr Pharm Des.* 2010;16:2298–2307.

60. Shilo M, Berenstein P, Dreifuss T, et al. Insulin-coated gold nanoparticles as a new concept for personalized and adjustable glucose regulation. *Nanoscale.* 2015a;7:20489–20496.

61. Shilo M, Sharon A, Baranes K, Motiei M, Lellouche JPM, Popovtzer R. The effect of nanoparticle size on the probability to cross the blood-brain barrier: an *in-vitro* endothelial cell model. *J Nanobiotechnol.* 2015b;13:19.

62. Zhang G, Yang Z, Lu W, et al. Influence of anchoring ligands and particle size on the colloidal stability and *in vivo* biodistribution of polyethylene glycol coated gold nanoparticles in tumor-xenografted mice. *Biomaterials.* 2009;30:1928–1936.

63. Barbu E, Molnar E, Tsibouklis J, Gorecki DC. The potential for nanoparticle-based drug delivery to the brain:

overcoming the blood–brain barrier. *Expert Opin Drug Deliv.* 2009;6:553–565.

64. Gao K, Jiang X. Influence of particle size on transport of methotrexate across blood–brain barrier by polysorbate 80-coated polybutylcyanoacrylate nanoparticles. *Int J Pharm.* 2006;310:213–219.

65. Soni S, Babbar AK, Sharma RK, Banerjee T, Maitra A. Pharmacoscintigraphic evaluation of polysorbate 80-coated chitosan nanoparticles for brain targeting. *Am J Drug Deliv.* 2005;3:205–212.

65a. Fonseca-Santos B, Gremiao MP, Chorilli M. Nanotechnology-based drug delivery systems for the treatment of Alzheimer's disease. *Int J Nanomedicine.* 2015;10:4981–5003.

66. Johnsen KB, Moos T. Revisiting nanoparticle technology for blood-brain barrier transport: unfolding at the endothelial gate improves the fate of transferring receptor-targeted liposomes. *J Contr Rel.* 2016;222:32–46.

67. Gaillard Peter J, V CC, B.M D. Drug delivery to the brain – physiological concepts, methodologies and approaches. In: Hammarlund-Udenaes M, Lange ECMD, Thorne RG, eds. *Blood-to-Brain Drug Delivery Using Nanocarriers.* New York: Springer; 2014:433–454.

68. Jain KK. Nanobiotechnology-based drug delivery to the central nervous system. *Neurodegener Dis.* 2007;4:287–291.

69. Jain KK. Nanoneurology. In: *Applications of Biotechnology in Neurology.* New York: Humana Press; 2013:283–294.

70. Brambilla D, Verpillot R, De Kimpe L, Taverna M, Le Droumaguet B, Andrieux K. Nanoparticles against Alzheimer's disease: PEG-PACA nanoparticles are able to link the Aβ-peptide and influence its aggregation kinetic. *J Biotechnol.* 2010a;150:27.

71. Brambilla D, Verpillot R, Taverna M, et al. New method based on capillary electrophoresis with laser-induced fluorescence detection (CE-LIF) to monitor interaction between nanoparticles and the amyloid-β peptide. *Anal Chem.* 2010b;82:10083–10089.

72. Aalinkeel R, Kutscher HL, Singh A, et al. Neuroprotective effects of a biodegradable poly(lactic-co-glycolic acid)-ginsenoside Rg3 nanoformulation: a potential nanotherapy for Alzheimer's disease? *J Drug Target.* 2017;17:1–12.

73. Liu Y, An S, Li J, et al. Brain-targeted co-delivery of therapeutic gene and peptide by multifunctional nanoparticles in Alzheimer's disease mice. *Biomaterials.* 2016;80:33–45.

74. Wilson B, Samanta MK, Santhi K, Kumar KP, Paramakrishnan N, Suresh B. Poly(n-butylcyanoacrylate) nanoparticles coated with polysorbate 80 for the targeted delivery of rivastigmine into the brain to treat Alzheimer's disease. *Brain Res.* 2008;1200:159–168.

75. Sun W, Xie C, Wang H, Hu Y. Specific role of polysorbate 80 coating on the targeting of nanoparticles to the brain. *Biomaterials.* 2004;25:3065–3071.

76. Amorim Cde M, Couto AG, Netz DJ, de Freitas RA, Bresolin TM. Antioxidant idebenone-loaded nanoparticles based on chitosan and N-carboxymethylchitosan. *Nanomedicine.* 2010;6:745–752.

77. Mattson MP. Pathways towards and away from Alzheimer's disease. *Nature.* 2004;430:631–639.

78. Sinha M, Bhowmick P, Banerjee A, Chakrabarti S. Antioxidant role of amyloid β protein in cell-free and biological systems: implication for the pathogenesis of Alzheimer disease. *Free Radic Biol Med.* 2013;56:184–192.

79. Wagner S, Zensi A, Wien SL, et al. Uptake mechanism of ApoE-modified nanoparticles on brain capillary endothelial cells as a blood-brain barrier model. *PLoS One.* 2012;7(3):e32568.

80. Lannfelt L, Relkin NR, Siemers ER. Amyloid-ß-directed immunotherapy for Alzheimer's disease. *J Intern Med.* 2014;275:284–295.

81. Carradori D, Gaudin A, Brambilla D, Andrieux K. Application of nanomedicine to the CNS diseases. *Int Rev Neurobiol.* 2016;130:73–113.

82. Guccione C, Oufir M, Piazzini V, et al. Andrographolide-loaded nanoparticles for brain delivery: formulation, characterisation and *in vitro* permeability using hCMEC/D3 cell line. *Eur J Pharm Biopharm.* 2017;119:253–263.

83. Lin HC, Ho MY, Tsen CM, et al. Comparative proteomics reveals silver nanoparticles alter fatty acid metabolism and amyloid beta clearance for neuronal apoptosis in a triple cell co-culture model of the blood-brain barrier. *Toxicol Sci.* 2017;158:151–163.

84. Shilo M, Motiei M, Hana P, Popovtzer R. Transport of nanoparticles through the blood–brain barrier for imaging and therapeutic applications. *Nanoscale.* 2014;6:2146–2152.

85. Meir R, Betzer O, Motiei M, Kronfeld N, Brodie C, Popovtzer R. Design principles for noninvasive, longitudinal and quantitative cell tracking with nanoparticle-based CT imaging. *Nanomedicine.* 2017;13:421–429.

86. Betzer O, Meir R, Dreifuss T, et al. *In-vitro* optimization of nanoparticle-cell labeling protocols for *in-vivo* cell tracking applications. *Sci Rep.* 2015;5:15400.

87. Rip J, Chen L, Hartman R, et al. Glutathione PEGylated liposomes: pharmacokinetics and delivery of cargo across the blood-brain barrier in rats. *J Drug Target.* 2014;22:460–467.

88. Do TD, Ul Amin F, Noh Y, Kim MO, Yoon J. Guidance of magnetic nanocontainers for treating Alzheimer's disease using an electromagnetic, targeted drug-delivery actuator. *J Biomed Nanotechnol.* 2016;12:569–574.

89. Ansari SA, Satar R, Perveen A, Ashraf GM. Current opinion in Alzheimer's disease therapy by nanotechnology-based approaches. *Curr Opin Psychiatry.* 2017;30:128–135.

90. Amin FU, Hoshiar AK, Do TD, et al. Osmotin-loaded magnetic nanoparticles with electromagnetic guidance for the treatment of Alzheimer's disease. *Nanoscale.* 2017;9:10619–10632.

91. Fiandra L, Mazzucchelli S, Truffi M, et al. *In vitro* Permeation of FITC-loaded Ferritins Across a Rat Blood-brain Barrier: a Model to Study the Delivery of Nanoformulated Molecules. *J Vis Exp.* 2016;114.

CHAPTER 6

Gene Therapy: The Cornerstone in the Development of Alzheimer's Disease Therapeutics

ABSTRACT

Multifarious genes play a pivotal role in Alzheimer's disease (AD) pathology. However, growing lines of evidence also shows that modulation or engineering of some genes offer pronounced protective effect against AD. This chapter primarily focuses on recent advancements in numerous gene therapies such as gene targeting, gene editing, role of heat shock proteins, cytokines, and their pros and cons. The remarkable efficacy of a multitarget small molecule, ASS234 in combating neuroinflammation by regulating multifarious genes and their expression has been discussed. It also throws light on a few shortcomings associated with gene therapy and future perspectives.

KEYWORDS

Alzheimer's disease; CRISPR/Cas9 system; Gene therapy; Heat shock protein; miRNA.

INTRODUCTION

Genes involved in Alzheimer's disease (AD) include amyloid precursor protein (APP), presenilin 1 (PS1), Apolipoprotein E (ApoE).[1–4] More recently, CRISPR/Cas9 system was employed to perturb mutant allele of *APPswe* mutation.[1] Accordingly, guide RNA (gRNA) was prepared and targeted against APP sw and APP WT. Deteriorated Aβ production was accomplished by this system. Presence of this allele within the proximity of β facilitated the recognition of gene appropriately. The primary strategy of this study is to employ pruned gRNA which can mitigate the complementarity length between gRNA and target DNA site and accomplish specific recognition.[1,5]

PRESENILIN

PS1 and PS2 play a pivotal role in $^\gamma$ secretase activity and intracellular Ca^{2+} signaling thus becoming the potential therapeutic targets for AD.[6] While the intracellular accumulation of $^\gamma$ secretase plays a key role in Aβ generation, the protein localization regulates intracellular trafficking of $^\gamma$ secretase.[6] Therefore, augmented accumulation of protein in endoplasmic reticulum mitigates mature APP thus curtailing surface accumulation.[6,7] Although deactivation of PS in the hippocampus has no role in endoplasmic reticulum (ER) Ca^{2+} concentration, ryanodine receptor levels and activity were diminished in the hippocampus in the absence of PS.[6] This forms the basis to understand the association between PS and calcium homeostasis and supports the notion that Ca^{2+} dyshomeostasis is an early pathological event in AD etiology.[6,8] Lack of Aβ plaque influence on hippocampal pyramidal neuron Ca^{2+} signaling emphasizes that Ca^{2+} signaling dysfunction is not Aβ dependent.[6,9] Lysosomal acidification in AD was found to be triggered by PS1 deletion, which is a prominent molecular underpinning and plays a role in early disease detection.[6,10]

STEM CELLS

Bone marrow–isolated mesenchymal stem cells (BM-MSCs) that are abundantly available offer remarkable low immunogenicity.[11,12] They have the ability to translocate to lesion tissues and secrete various neurotrophic growth factors to initiate regeneration and neuroprotection.[11,13–15] The better spanning of blood-brain barrier (BBB) and localization in brain was accomplished by MSCs after intracerebral, intracarotid, or intravenous injections, thus making them promising therapeutics in AD.[11,12] Accordingly, these cells induced neurogenesis and ameliorated learning and memory in animals by mitigating Aβ accumulation and alteration of microglial cells.[11,16–18]

Despite remarkable therapeutic effect of MSCs in AD treatment, their decreased survival rate due to the presence of high ROS, cytotoxic Aβ oligomers, and p-tau hindered their success.[10,11,19] With a view to overcoming this biological, riddle Han et al.[11] for the first time employed miRNA (microRNA (ribonucleic acid)) let-7f-5p to accrue MSCs' longevity. Surprisingly, these let-7f-5p showed an antiapoptotic role in Aβ$_{25-35}$ treated cells by modulating caspase-3 activity in AD models.[11] Caspase-3 has been corroborated as a potential target gene of let-7f-5p. While the primary role of MSCs is to substitute the impaired cells, they were also found to serve a few more roles such as mitigating amyloid accumulation, regulating inflammation, and enhancing endogenous neurogenesis in AD.[11,20]

MICRORNA/CDK5R1

Cyclin-dependent kinase five regulatory subunit 1 (CDK5R1) gene encodes for *p35*, the major activator of cyclin-dependent kinase 5 (CDK5). The *p35*/CDK5 complex plays a pivotal role in a plethora of events like brain development and function.[21] However, mounting evidence unraveled that impairment of this complex is linked to AD onset.[21,22] miRNA 15/107 family of microRNAs were found to be downregulated in the hippocampus and cortex of AD patients, whereas CDK5R1 mRNA levels were upregulated in the hippocampus.[21,23] In another study, neurons transfected with miRNA 200b/200c curtailed Aβ secretion.[24] It was found that miRNA 200c attenuated ribosomal protein S6 kinase B1 (S6K1), which phosphorylates insulin receptor substrate 1 at serine residues (IRS-1P Ser). In line with this, S6K1-dependent IRS-1P Ser inhibited insulin signaling that is often found in AD pathology.[24]

Long noncoding RNAs (lncRNAs) with more than 200 bases without protein coding ability add an extra layer of complexity to CDK5R1 expression control, which plays an essential role in AD pathogenesis.[21,25] Although lncRNAs' role in malignancies and as a diagnostic and prognostic marker has been corroborated long ago, their essential role in neurodegenerative diseases is recently unearthed.[21,26] Panoply of lncRNAs implicated in AD is BACE1-AS, 17A, and NDM29.[21,27] Among the lncRNAs studied, NEAT1 and HOTAIR negatively modulate CDK5R1 mRNA levels, but MALAT1 has a positive effect on CDK5R1 expression.[21] Therefore,

lncRNA mediating another level of CDK5R1 has influence on AD research.[21]

HEAT SHOCK PROTEIN

Heat shock proteins (HSPs) controlled by genes also contribute for protein misfolding and preparation of immunogenic proteins like HSP which mediate amyloid β oligomer accumulation.[28–33] The detection of heat shock gene Sirtuin 1 (Sirt 1), an NAD⁺ reliant class II histone deacetylase entailed in deacetylation of heat shock factors 1 (HSF1), protects neurons from proteotoxicity in chronic neurodegenerative disorders.[33–35] Sirt 1 being a calorie-sensitive gene gets suppressed by excess nutrition and bacterial lipopolysaccharide, thus influencing marginal amyloid β clearance pathway (Martins, 2018).[36–39]

BECLIN

An autophagy-associated protein Beclin 1 has been found to be deteriorated in AD, thus mediating Aβ aggregation in mice.[40] Studies on APP+Becn1+/− showed that Beclin one deficiency induced change in microglial response to Aβ, decreased autophagy, augmented extra- and intracellular Aβ accumulation synaptodendritic diminution, and neuronal damage.[40,41] Studies have shown that genetically modified fibroblasts grafted into the forebrain of AD patients profoundly ameliorated cognition without induction of any side effects.[40,42]

APOLIPOPROTEIN E

Apolipoprotein E (Apo E) which is mainly generated in the liver profoundly expressed in the brain plays a vital role in physiological activities such as neuronal plasticity, synaptogenesis, Aβ fibril formation and accumulation, degeneration and scavenging of Aβ.[40,43–47] Among the three ApoE isoforms with the difference of only 2 residues ApoE 2 offers neuroprotection, ApoE 3 is neutral, whereas ApoE4 is the vital genetic risk factor for AD. Several lines of evidence also showed association between ApoE and tau.[40,48]

DRUG MECHANISM/GENE MODIFICATION

More recently, several lines of evidence have shown that ceftriaxone, an antibiotic with adequate neuroprotective efficacy, exhibited a profound influence on genes. In line with this, ceftriaxone showed remarkable reduction in mRNA levels of *Bace1* (encode) and *Ace2* in the hypothalamus and Aktb in the frontal cortex in OXYS rats.[49] Additionally, enhanced Mme, Ide, and Epo mRNA levels in amygdala and the levels of Ece1 and Aktb in the striatum were also observed.[49] Since Aβ formation depends on APP cleavage by BACE 1 and its scavenging by mechanisms such as proteolytic breakdown, transport, and aggregation, the corresponding gene modulation strategies were employed.[49-53]

NEUROINFLAMMATION AND CYTOKINES

Neuroinflammation plays a pivotal role in AD pathology.[54] Therefore, a plethora of therapies targeted at neuroinflammation hold considerable promise currently and attract the attention of neuroscientists.[55-58]

Nevertheless, there is also considerable dearth in neuroinflammation-targeted therapies. To overcome this, a robust BBB spanning compound N-((5-(3-(1-benzylpiperidin-4-yl)propoxy)-1-methyl-1H-indol-2-yl)methyl)-N-methylprop-2-yn-1-amine (ASS234) was studied adequately.[59-66] However, no adequate studies were found on the toxicity profile of this compound. To this end the influence on genes was studied to abridge the gap. ASS234 successfully suppressed the inflammation in lipopolysaccharide (LPS)-instigated RAW 264.7 macrophages. In order to understand the underlying mechanism of neuroprotection gene expression of IL-6, IL1β, TNF-α, TNFR1, NF-κB, IL-10, and TGF-β was studied in SH-SY5Y cells.[57,67-70] Because they play an essential role in induction of inflammatory response, their regulation is an appropriate avenue to combat inflammation.[57] NF-κB was found to control IL-6, TNF-α, and IL-1β, which are responsible for BBB impairment.[57] ASS234 being a robust suppressor of this mechanistic pathway curtailed BBB impairment in AD patients.[57] ASS234 treatment also upregulated IL-10 and TGF-β in turn initiated the antiapoptotic pathway and augmented cell survival.[57]

An array of studies has shown that deteriorated interleukin-33 (IL-33) transcript and protein levels were observed in AD patients unlike in healthy individuals.[40,71,72] Introduction of IL-33 into APP/PS1 transgenic mice showed amelioration in synaptic plasticity and cognitive deficit.[40,72] In addition, IL-33 treatment induces an improvement in the anti-inflammatory genes such as Arg 1 and Fizz 1 in the microglia and the macrophages in the brain, initiating a varied activation state with improved Aβ phagocytosis.[40]

CONCLUSION

Despite significant contribution of gene therapy in progress of AD therapeutics, a few limitations are to be addressed. Unfortunately, manifold immunotherapeutic studies targeted against Aβ yielded unsuccessful results.[1,73,74] In spite of remarkable therapeutic efficacy of a few strategies targeted against neuroinflammation in AD, lack of molecular understanding has been a hurdle in making them robust therapeutic arsenals. To overcome these issues a multitargeted small molecule, ASS234, synthesized and extensively used to downregulate inflammatory genes holds sufficient promise in AD therapeutics currently.[57] Therefore, there is a burgeoning need to show much focus on this compound to design it as a clinically appropriate tool to combat AD.

REFERENCES

1. Gyorgy B, Loov C, Zaborowski MP, et al. CRISPR/Cas9 mediated disruption of the Swedish APP allele as a therapeutic approach for early-onset Alzheimer's disease. *Mol Ther Nucleic Acids*. 2018;11:429–440.
2. Obulesu M, Somashekhar R, Venu R. Genetics of alzheimer's disease: an insight into presenilins and apolipoprotein E instigated neurodegeneration. *Int J Neurosci*. 2011;121(5):229–236.
3. Bertram L, Tanzi RE. The genetics of alzheimer's disease. *Prog Mol Biol Transl Sci*. 2012;107:79–100.
4. Bertram L, Tanzi RE. Genome-wide association studies in Alzheimer's disease. *Hum Mol Genet*. 2009;18(R2):R137–R145.
5. Fu Y, Sander JD, Reyon D, Cascio VM, Joung JK. Improving CRISPR-Cas nuclease specificity using truncated guide RNAs. *Nat Biotechnol*. 2014;32:279–284.
6. Scott Duncan R, Song B, Peter K. Presenilins as drug targets for Alzheimer's disease—recent insights from cell biology and electrophysiology as novel opportunities in drug development. *Int J Mol Sci*. 2018;19:1621.
7. Park HJ, Shabashvili D, Nekorchuk MD, et al. Retention in endoplasmic reticulum 1 (RER1) modulates amyloid-(Aβ) production by altering trafficking of -secretase and amyloid precursor protein (APP). *J Biol Chem*. 2012;287:40629–40640.
8. Wu B, Yamaguchi H, Lai FA, Shen J. Presenilins regulate calcium homeostasis and presynaptic function via ryanodine receptors in hippocampal neurons. *Proc Natl Acad Sci USA*. 2013;110:15091–15096.
9. Briggs CA, Schneider C, Richardson JC, Stutzmann GE. β amyloid peptide plaques fail to alter evoked neuronal calcium signals in APP/PS1 Alzheimer's disease mice. *Neurobiol Aging*. 2013;34:1632–1643.
10. Lee JH, McBrayer MK, Wolfe DM, et al. Presenilin 1 maintains lysosomal Ca²⁺ homeostasis via TRPML1 by regulating vATPase-mediated lysosome acidification. *Cell Rep*. 2015;12:1430–1444.

11. Han L, Zhou Y, Zhang R, Wu K, Lu Y, Li Y. MicroRNA let-7f-5p promotes bone marrow mesenchymal stem cells survival by targeting caspase-3 in Alzheimer disease model. *Front Neurosci.* 2018;12:333.

12. Lo Furno D, Mannino G, Giuffrida R. Functional role of mesenchymal stem cells in the treatment of chronic neurodegenerative diseases. *J Cell Physiol.* 2017;233:3982–3999.

13. De Becker A, Riet IV. Homing and migration of mesenchymal stromal cells: how to improve the efficacy of cell therapy. *World J Stem Cell.* 2016;8:73–87.

14. Koniusz S, Andrzejewska A, Muraca M, Srivastava AK, Janowski M, Lukomska B. Extracellular vesicles in physiology, pathology, and therapy of the immune and central nervous system, with focus on extracellular vesicles derived from mesenchymal stem cells as therapeutic tools. *Front Cell Neurosci.* 2016;10:109.

15. Petrou P, Gothelf Y, Argov Z, et al. Safety and clinical effects of mesenchymal stem cells secreting neurotrophic factor transplantation in patients with amyotrophic lateral sclerosis: results of phase 1/2 and 2a clinical trials. *JAMA Neurol.* 2016;73:337–344.

16. Duncan T, Valenzuela M. Alzheimer's disease, dementia, and stem cell therapy. *Stem Cell Res Ther.* 2017;8:111.

17. Liew LC, Katsuda T, Gailhouste L, Nakagama H, Ochiya T. Mesenchymal stem cell-derived extracellular vesicles: a glimmer of hope in treating Alzheimer's disease. *Int Immunol.* 2017;29:11–19.

18. Oh SH, Kim HN, Park HJ, Shin JY, Lee PH. Mesenchymal stem cells increase hippocampal neurogenesis and neuronal differentiation by enhancing the Wnt signaling pathway in an Alzheimer's disease model. *Cell Transplant.* 2015;24:1097–1109.

19. Oh S, Son M, Choi J, Lee S, Byun K. sRAGE prolonged stem cell survival and suppressed RAGE-related inflammatory cell and T lymphocyte accumulations in an Alzheimer's disease model. *Biochem Biophys Res Commun.* 2018;495:807–813.

20. Kim JY, Kim DH, Kim JH, et al. Soluble intracellular adhesion molecule-1 secreted by human umbilical cord blood-derived mesenchymal stem cell reduces amyloid-b plaques. *Cell Death Differ.* 2012;19:680–691.

21. Spreafico M, Grillo B, Rusconi F, et al. Multiple layers of *CDK5R1* Regulation in Alzheimer's Disease Implicate Long Non-Coding RNAs. *Int J Mol Sci.* 2018;19. (pii: E2022).

22. Wang JZ, Grundke-Iqbal I, Iqbal K. Kinases and phosphatases and tau sites involved in Alzheimer neurofibrillary degeneration. *Eur J Neurosci.* 2007;25:59–68.

23. Moncini S, Lunghi M, Valmadre A, et al. The miR-15/107 family of microRNA genes regulates CDK5R1/p35 with implications for Alzheimer's disease pathogenesis. *Mol Neurobiol.* 2017;54:4329–4342.

24. Higaki S, Muramatsu M, Matsuda A, et al. Defensive effect of microRNA-200b/c against amyloid-beta peptide-induced toxicity in Alzheimer's disease models. *PLoS One.* 2018;13:e0196929.

25. Khorkova O, Hsiao J, Wahlestedt C. Basic biology and therapeutic implications of lncRNA. *Adv Drug Deliv Rev.* 2015;87:15–24.

26. Li Y, Wang X. Role of long noncoding RNAs in malignant disease (Review). *Mol Med Rep.* 2016;13:1463–1469.

27. Faghihi MA, Kocerha J, Modarresi F, et al. RNAi screen indicates widespread biological function for human natural antisense transcripts. *PLoS One.* 2010;5. pii: e13177.

28. Canto C. The heat shock factor HSF1 juggles protein quality control and metabolic regulation. *J Cell Biol.* 2017;216.

29. Dalgediene I, Lasickiene R, Budvytyte R, Valincius G, Morkuniene R, Borutaite V. Immunogenic properties of amyloid beta oligomers. *J Biomed Sci.* 2013;20:10.

30. Gomez-Pastor R, Burchfiel ET, Thiele DJ. Regulation of heat shock transcription factors and their roles in physiology and disease. *Nat Rev Mol Cell Biol.* 2018;19:4–19.

31. Maier M, Seabrook TJ, Lazo ND, et al. Short amyloid-beta (Abeta) immunogens reduce cerebral Abeta load and learning deficits in an Alzheimer's disease mouse model in the absence of an Abeta-specific cellular immune response. *J Neurosci.* 2006;26:4717–4728.

32. Vabulas RM, Raychaudhuri S, Hayer-Hartl M, Ulrich F. Protein folding in the cytoplasm and the heat shock response. *Cold Spring Harb Perspect Biol.* 2010;2:a004390.

33. Martins IJ. Heat shock gene inactivation and protein aggregation with links to chronic diseases. 2018;6(2). pii:E39.

34. Martins IJ. Heat shock gene Sirtuin 1 regulates post-prandial lipid metabolism with relevance to nutrition and appetite regulation in diabetes. *Int J Diabetes Clin Diagn.* 2016;3:20.

35. Martins IJ. Type 3 diabetes with links to NAFLD and other chronic diseases in the western world. *Int J Diabetes.* 2016;1:1–5.

36. Martins IJ. Overnutrition determines LPS regulation of mycotoxin induced neurotoxicity in neurodegenerative diseases. *Int J Mol Sci.* 2015;16:29554–29573.

37. Martins IJ. Calorie sensitive anti-aging gene regulates hepatic amyloid beta clearance in diabetes and neurodegenerative diseases. *EC Nutr.* 2017;1:30–32.

38. Martins IJ. Unhealthy diets determine benign or toxic amyloid beta states and promote brain amyloid beta aggregation. *Austin J Clin Neurol.* 2015;2:1060–1066.

39. Martins IJ. The future of genomic medicine involves the maintenance of Sirtuin 1 in global populations. *Int J Mol Biol.* 2017;2:00013.

40. Raikwar SP, Ramasamy T, Iuliia D, Mohammad Ejaz A, Govindhasamy Pushpavathi S. Neuro-immuno-gene- and genome-editing-therapy for Alzheimer's disease: are we there yet? *J Alzheimers Dis.* 2018;65:321–344.

41. Pickford F, Masliah E, Britschgi M, Lucin K, Narasimhan R, Jaeger PA. The autophagy-related protein beclin 1 shows reduced expression in early Alzheimer disease and regulates amyloid beta accumulation in mice. *J Clin Invest.* 2008;118:2190–2199.

42. Tuszynski MH, Thal L, Pay M, Salmon DP, U HS, Bakay R. A phase 1 clinical trial of nerve growth factor gene therapy for Alzheimer disease. *Nat Med.* 2005;11:551–555.

43. Corder EH, Saunders AM, Risch NJ, et al. Protective effect of apolipoprotein E type 2 allele for late onset Alzheimer disease. *Nat Genet.* 1994;7:180–184.

44. Fagan AM, Bu G, Sun Y, Daugherty A, Holtzman DM. Apolipoprotein E-containing high density lipoprotein promotes neurite outgrowth and is a ligand for the low density lipoprotein receptor-related protein. *J Biol Chem.* 1996;271:30121–30125.

45. Holtzman DM, Bales KR, Tenkova T, et al. Apolipoprotein E isoform dependent amyloid deposition and neuritic degeneration in a mouse model of Alzheimer's disease. *Proc Natl Acad Sci USA.* 2000a;97:2892–2897.

46. Holtzman DM, Fagan AM, Mackey B, Tenkova T, Sartorius L, Paul SM. Apolipoprotein E facilitates neuritic and cerebrovascular plaque formation in an Alzheimer's disease model. *Ann Neurol.* 2000b;47:739–747.

47. Mahley RW. Apolipoprotein E: cholesterol transport protein with expanding role in cell biology. *Science.* 1988;240:622–630.

48. Shi Y, Yamada K, Liddelow SA, et al. ApoE4 markedly exacerbates tau-mediated neurodegeneration in a mouse model of tauopathy. *Nature.* 2017;549:523–527.

49. Tikhonova MA, Amstislavskaya TG, Belichenko1 VM, et al. Modulation of the expression of genes related to the system of amyloid-beta metabolism in the brain as a novel mechanism of ceftriaxone neuroprotective properties. *BMC Neurosci.* 2018;19:13.

50. Grimm MO, Mett J, Stahlmann CP, Haupenthal VJ, Zimmer VC, Hartmann T. Neprilysin and Abeta clearance: impact of the APP intracellular domain in NEP regulation and implications in Alzheimer's disease. *Front Aging Neurosci.* 2013;5:98.

51. Nalivaeva NN, Belyaev ND, Turner AJ. New insights into epigenetic and pharmacological regulation of amyloid-degrading enzymes. *Neurochem Res.* 2016;41:620–630.

52. Singh SK, Srivastav S, Yadav AK, Srikrishna S, Perry G. Overview of Alzheimer's disease and some therapeutic approaches targeting Abeta by using several synthetic and herbal compounds. *Oxid Med Cell Longev.* 2016;2016:7361613.

53. Wang DS, Dickson DW, Malter JS. Beta-amyloid degradation and Alzheimer's disease. *J Biomed Biotechnol.* 2006;2006:58406.

54. Obulesu M, Jhansilakshmi M. Neuroinflammation in Alzheimer's disease: an understanding of physiology and pathology. *Int J Neurosci.* 2014;124:227–235.

55. Ardura-Fabregat A, Boddeke E, Boza-Serrano A, et al. Targeting neuroinflammation to treat Alzheimer's disease. *CNS Drugs.* 2017;31:1057–1082.

56. Dansokho C, Heneka MT. Neuroinflammatory responses in Alzheimer's disease. *J Neural Transm.* 2018;125:771.

57. del Pino J, Marco-Contelles J, Lopez-Munoz F, Romero A, Ramos E. Neuroinflammation signaling modulated by ASS234 a multitarget small molecule for Alzheimer's disease therapy. *ACS Chem Neurosci.* 2018 (in press).

58. Lyman M, Lloyd DG, Ji X, Vizcaychipi MP, Ma D. Neuroinflammation: the role and consequences. *Neurosci Res.* 2014;79:1–12.

59. Bolea I, Gella A, Monjas L, et al. Multipotent, permeable drug ASS234 inhibits Abeta aggregation, possesses antioxidant properties and protects from Abeta-induced apoptosis *in vitro. Curr Alzheimer Res.* 2013;10:797–808.

60. Bolea I, Juarez-Jimenez J, de Los Rios C, et al. Synthesis, biological evaluation, and molecular modeling of donepezil and N-[(5-(benzyloxy)-1-methyl-1H-indol-2-yl) methyl]-N-methylprop-2-yn-1-amine hybrids as new multipotent cholinesterase/monoamine oxidase inhibitors for the treatment of Alzheimer's disease. *J Med Chem.* 2011;54:8251–8270.

61. del Pino J, Ramos E, Aguilera OM, Marco-Contelles J, Romero A. Wnt signaling pathway, a potential target for Alzheimer's disease treatment, is activated by a novel multitarget compound ASS234. *CNS Neurosci Ther.* 2014;20:568–570.

62. Esteban G, Van Schoors J, Sun P, et al. In-vitro and in-vivo evaluation of the modulatory effects of the multitarget compound ASS234 on the monoaminergic system. *J Pharm Pharmacol.* 2017;69:314–324.

63. Esteban G, Allan J, Samadi A, et al. Kinetic and structural analysis of the irreversible inhibition of human monoamine oxidases by ASS234, a multi-target compound designed for use in Alzheimer's disease. *Biochim Biophys Acta Protein Proteonomics.* 2014;1844:1104–1110.

64. Marco-Contelles J, Unzeta M, Bolea I, et al. ASS234, as a new multi-target directed propargylamine for Alzheimer's disease therapy. *Front Neurosci.* 2016;10:294.

65. Ramos E, Romero A, Marco-Contelles J, Del Pino J. Upregulation of antioxidant enzymes by ASS234, a multitarget directed propargylamine for Alzheimer's disease therapy. *CNS Neurosci Ther.* 2016;22:799–802.

66. Serrano MP, Herrero-Labrador R, Futch HS, Serrano J, Romero A, Fernandez AP. The proof-of-concept of ASS234: peripherally administered ASS234 enters the central nervous system and reduces pathology in a male mouse model of Alzheimer disease. *J Psychiatry Neurosci.* 2017;42:59–69.

67. Medeiros R, Prediger RD, Passos GF, et al. Connecting TNF-alpha signaling pathways to iNOS expression in a mouse model of Alzheimer's disease: relevance for the behavioral and synaptic deficits induced by amyloid beta protein. *J Neurosci.* 2007;27:5394–53404.

68. Siren AL, McCarron R, Wang L, Garcia-Pinto P, Ruetzler C, Martin D. Proinflammatory cytokine expression contributes to brain injury provoked by chronic monocyte activation. *Mol Med.* 2001;7:219–229.

69. Su F, Bai F, Zhang Z. Inflammatory cytokines and Alzheimer's disease: a review from the perspective of genetic polymorphisms. *Neurosci Bull.* 2016;32:469–480.

70. Zheng Y, Fang W, Fan S, Liao W, Xiong Y, Liao S. Neurotropin inhibits neuroinflammation via suppressing NF-kappaB and MAPKs signaling pathways in lipopolysaccharide-stimulated BV2 cells. *J Pharmacol Sci.* 2018;136:242.

71. Chapuis J, Hot D, Hansmannel F, et al. Transcriptomic and genetic studies identify IL-33 as a candidate gene for Alzheimer's disease. *Mol Psychiatr.* 2009;14:1004–1016.

72. Fu AK, Hung KW, Yuen MY, Zhou X, Mak DS, Chan IC. IL-33 ameliorates Alzheimer's disease-like pathology and cognitive decline. *Proc Natl Acad Sci U S A*. 2016;113:E2705–E2713.

73. Doody RS, Thomas RG, Farlow M, et al. Phase 3 trials of solanezumab for mild-to-moderate Alzheimer's disease. *N Engl J Med*. 2014;370:311–321.

74. Salloway S, Sperling R, Fox NC, et al. Bapineuzumab 301 and 302 Clinical Trial Investigators. Two phase 3 trials of bapineuzumab in mild-to-moderate Alzheimer's disease. *N Engl J Med*. 2014;370:322–333.

CHAPTER 7

Viral Vector Therapeutics Against Alzheimer's Disease

ABSTRACT

Aβ peptide accumulation on extracellular surface has been one of the hallmarks of Alzheimer's disease (AD). To overcome this accumulation and protein misfolding, a plethora of therapeutics is currently in progress. Viral vector therapeutics which render appropriate amelioration of various disease symptoms were used in AD treatment also. More specifically, adeno-associated virus was extensively used and found intensely effective in overcoming biological riddles such as blood-brain barrier permeability. This chapter throws light on various viral vector therapeutics and their merits and demerits.

KEYWORDS

Adeno-associated virus; Alzheimer's disease; Neprilysin; Progranulin; Viral vectors.

INTRODUCTION

Viruses have long been in use as vectors due to their remarkably established mechanisms for effective and safe delivery of genetic material to the host cells.[1] Several transgenic animal models exhibited remarkably low Aβ levels after vaccination.[2-6] Despite significant efficacy, a few adverse effects were observed and clinical trials declined.[6-8] While a wide range of vaccines were employed in disease treatment in the past, viral vector therapeutics has been gaining grounds recently.[1] Viral vectors that deliver transgenes which code for counteracting antibodies to nonhematopoietic tissues can substantiate their long-term expression, thus aiding the treatment or prevention of copious infectious diseases.[1] These transgenes mediate the production of monoclonal antibodies (mAbs) in nonhematopoietic cells which release mAbs into circulation or the indigenous environment.[1]

Viral Vector Therapeutics

Wealth of studies reported that viral vectors are potential therapeutics compared to nanotechnology-based DDS because of their pronounced efficacy and safety.[9] In order to accomplish substantial neuroprotection against Alzheimer's disease (AD), viral vector expression of Aβ processing enzymes, such as neprilysin[10,11] or anti-Aβ single-chain antibodies, to achieve passive immunization[12] was widely used.[9] Additionally, regulation of amyloid precursor protein deterioration was achieved by genetic transfer of siRNA for β-secretase[9,13] or shRNA to curb G protein–coupled receptor for γ-secretase.[9,14] Multifarious other viral vector therapeutics comprise anti-Apo E antibodies,[15] inhibition of acylCoA-cholesterol acyltransferase,[16] distribution of growth factors such as cerebral dopamine neurotrophic factor,[17] insulin growth factor,[18] brain-derived neurotrophic factor,[19] and nerve growth factor[20] to avoid toxic injury to the brain. Inspite of merits of viral vector therapeutics, a few shortcomings such as immune intervention, absence of appropriate entry receptors, and inappropriate cellular uptake of vectors impede their success.[9]

Adeno-Associated Viral Vectors

Although multifarious expression platforms have been discovered, adeno-associated viral vectors were found to play a prominent role in disease therapeutics thus escalated to clinical translation.[1] This is currently considered an appreciable substitute to immunogen-based vaccines.[1] Substantial translation of these therapeutic maneuvers to patients may show appropriate ways to combat several diseases, including neurodegenerative diseases.[1] Naked plasmid DNA was found amenable for transfection due to their meek utilization, nonimmunogenicity, and feasibility of escalation to large-scale production.[1]

Adeno-associated virus (AAV) without relationship with disease in humans and recombinant vectors obtained from AAV offers stable gene expression without incorporation by making extrachromosomal concatamers of the distributed transgene sequences.[1,21] They are also nonpathogenic and nontoxic with a potential to show long-term expression even in nondividing cells like neurons.[6,22] The serotype embedded

Alzheimer's Disease Theranostics. https://doi.org/10.1016/B978-0-12-816412-9.00007-0
Copyright © 2019 Elsevier Inc. All rights reserved.

in the vector shows potential impact on its capability to transduce various tissues and also plays a substantial role in immunogenicity of the vector in a wide range of animal models.[1,21,23-25] In line with this, studies in murine transgenic AD models with Aβ expressing AAV showed remarkably reduced Aβ burden and cognitive deficits.[6,26] In another study, orally administered Aβ expressing AAV vector also showed reduced Aβ load in the brain.[6,27]

Recent studies have also shown in a comparative study between AAV9 and AAV2 that AAV9 with enhanced BBB spanning potential by an active, cell-mediated process and low transduction efficacy in brain microvascular endothelial cells (BMVECs) disintegrates inflammatory pathways.[28] Nevertheless, AAV2 showed better transgene expression in BMVEC cells.[28-34] Therefore, they are considered potent therapeutic tools and are presently in clinical trials. Surface receptors have been found to play an essential role in initiating enhanced cellular transduction by AAV2 compared to AAV9. AAV2 attaches to heparin sulfate proteoglycans, gets embedded in endosomal-like structures, trespasses to nucleus together with microtubule bundles, and gets localized in nucleus via nuclear pore complexes.[28,35-37]

Adenoviral vector is safe since it remains as an episome without incorporation into the chromosomes in eukaryotic cells.[38-40]

Wealth of studies has proven the clinical efficacy of AAV and amenable expression profile, thus making them vital tools for the delivery of mAbs *in vivo*.[1] Nevertheless, considerable low packaging ability of 5 kilobases impedes the expression of both heavy and light chain.[1] To circumvent this biological riddle, multifarious novel avenues were chosen such as smaller bivalent single-chain antibodies or immunoadhesins, chimeric antibody-like molecules, and heterologous viral sequences.[1,41-46]

mAbs

Since protein misfolding and aggregation has been the underlying cause of AD, mAbs that can locate misfolded proteins and curb their aggregation are currently in demand.[1] Single-chain antibodies (scFvs) were developed for Aβ, successfully mitigated Aβ accumulation, and ameliorated AD symptoms.[1,47] AAVs expressing genes of anti-Aβ mAbs showed remarkable protective efficacy by reducing Aβ accumulation and improving cognition in rodent AD models.[6,44,48-51]

In line with this, intrahippocampal injection of AAV1 encoding an Aβ-scFv showed reduced levels of insoluble Aβ, augmented microglia, and ameliorated cognition in a transgenic mouse model.[1,50] Since a

few studies showed adverse effects such as hemorrhaging after direct administration of AAV5 vectors into the brain, intramuscular (IM) injection of AAV2 expressing Aβ-scFvs was chosen as an alternative which profoundly mitigates physiologic and behavioral symptoms of AD.[1,44,45,52] In another study, IM injection of AAV1 expressing a full length Aβ-mAb exhibited controlled anti-Aβ levels above 100 μg/mL in serum which were retained for upto 64 weeks after injection. Accordingly, there is a possibility to observe low levels of Aβ in the brain.[1,51]

Tau

Growing lines of evidence showed that intrahippocampal introduction of AAV-vectored anti-phospho-tau antibody PHF1 results in ~50 folds increased antibodies in hippocampus compared to repeated systemic administration of anti-tau monoclonal antibodies.[53,54] Because of enhanced localization of antibodies against tau in the hippocampus, an 80%–90% decrease was observed in the levels of insoluble pathological tau species and neurofibrillary tangles in P301S tau transgenic mice.[54] Furthermore, in AAV anti-phospho-tau antibody PHF1 delivery, antibody delivery was preferably limited to neurons and increased antibody expression was localized in the cytoplasm of neurons.[54,55] Surprisingly, hippocampal volumes were not decreased in AAV-vectored antibody PHF1–treated P301S mice, whereas untreated mice showed around 8% decreased hippocampal volumes.[54] In line with this, extensive use of AAV-vectored genes was currently observed in mice via intrathalamic administration[56-58] and intrathecal[59] and systemic administration.[54,60,60a]

PROGRANULIN

Progranulin (PGRN), a multifunctional protein with profound expression in the brain, plays a pivotal role in mediating CNS neuroinflammation and acts as an autocrine neuronal growth factor.[61-63] Furthermore, it was also found to influence cell signaling pathways by controlling excitotoxicity, oxidative stress, synaptogenesis, and amyloid production.[63] Changes in PGRN levels are associated with AD pathology.[63-65] Additionally, a few studies emphasized the role of PGRN in Aβ accumulation, neuroinflammation, and toxicity.[63,66-68] Based on these findings, viral vector delivery of PGRN into transgenic mice Tg2576 showed profound expression in the hippocampus followed by significant reduction amyloid plaque burden, inflammatory markers, and synaptic atrophy.[63] PGRN has also been found to influence neuroinflammation via both anti-inflammatory and proinflammatory pathways.[63,69,70] PGRN attaches to TNFR (tumor necrosis factor receptor) and averts the binding of TNF-α to its receptors, which is considered the robust molecular underpinning in its anti-inflammatory activity.[63,71]

NEPRILYSIN

Neprilysin (NEP) profoundly disintegrates both monomeric and oligomeric forms.[63] Accordingly, NEP deterioration has also been implicated in AD pathology.[63] Therefore, NEP delivery and expression enhancement has been chosen as a potential AD therapeutic currently.[63,72] NEP reduction exhibited enhanced levels of Aβ40 and Aβ42 in APP transgenic mice, which were remarkably reduced by the delivery of NEP-expressing viral vectors.[63,73,74] Interestingly, augmented PGRN expression delivered through AAV vectors also ameliorates NEP expression, thus contributing for Aβ plaque reduction.[63]

GENE-BASED IMMUNOTHERAPEUTIC METHODS

Gene-based immunotherapeutic techniques are categorized into two types, namely, viral vector and nonviral vector (i.e., naked DNA) delivery methods.[6,75,76] Viral vector–based strategies are more effective compared to inactivated peptide/protein preparations or passive antibody introduction since the former requires only single administration while the latter needs repeated immunization.[6,77] Studies have shown that intranasal inoculation of adenovirus vaccine encoding Aβ 3–10 and CpG motif in BALB/c mouse is an effective immunotherapy against AD.[40] Adeno viral vectors probably translocate into CNS through retrograde transport exhibited by olfactory neurons of nasal epithelium.[40,78]

CONCLUSION

Despite remarkable progress in viral vector therapeutics against AD, their potential needs to be corroborated.[6] Although viral vector therapeutics improved symptoms of several diseases, yet a few demerits such as induction of autoimmunity and initiation of preexisting immunity against viral vectors limit their success.[6] Although AAV9 has been found to use galactose binding and absorptive endocytosis to span BBB, yet further studies are warranted to corroborate these mechanisms.[28]

REFERENCES

1. Deal CE, Balazs AB. Engineering humoral immunity as prophylaxis or therapy. *Curr Opin Immunol.* 2015;35:113–122.
2. Morgan D, Diamond DM, Gottschall PE, et al. A beta peptide vaccination prevents memory loss in an animal model of Alzheimer's disease. *Nature.* 2000;408:982–985.
3. Schenk D, Barbour R, Dunn W, et al. Immunization with amyloid-beta attenuates Alzheimer-disease-like pathology in the PDAPP mouse. *Nature.* 1999;400:173–177.
4. Lambracht-Washington D, Rosenberg RN. Advances in the development of vaccines for Alzheimer's disease. *Discov Med.* 2013;15:319–326.
5. Jindal H, Bhatt B, Sk S, Singh Malik J. Alzheimer disease immunotherapeutics: then and now. *Hum Vaccines Immunother.* 2014;10:2741–2743.
6. Kudrna JJ, Ugen KE. Gene-based vaccines and immunotherapeutic strategies against neurodegenerative diseases: potential utility and limitations. *Hum Vaccines Immunother.* 2015;11(8):1921–1926.
7. Gilman S, Koller M, Black RS, et al. Clinical effects of Abeta immunization (AN1792) in patients with AD in an interrupted trial. *Neurology.* 2005;64:1553–1562.
8. Holmes C, Boche D, Wilkinson D, et al. Long-term effects of Abeta42 immunisation in Alzheimer's disease: follow-up of a randomised, placebo-controlled phase I trial. *Lancet.* 2008;372:216–223.
9. Choudhury SR, Hudry E, Maguire CA, Sena-Esteves M, Breakefield XO, Grandi P. Viral vectors for therapy of neurologic diseases. *Neuropharmacology.* 2016;120:63–80.
10. Li Y, Wang J, Zhang S, Liu Z. Neprilysin gene transfer: a promising therapeutic approach for Alzheimer's disease. *J Neurosci Res.* 2015;93:1325–1329.
11. Lebson L, Nash K, Kamath S, et al. Trafficking CD11b-positive blood cells deliver therapeutic genes to the brain of amyloid-depositing transgenic mice. *J Neurosci.* 2010;30:9651–9658.
12. Levites Y, O'Nuallain B, Puligedda RD, et al. A human monoclonal IgG that binds Aβ assemblies and diverse amyloids exhibits anti-amyloid activities *in vitro* and *in vivo. J Neurosci.* 2015;35:6265–6276.
13. Nawrot B. Targeting BACE with small inhibitory nucleic acids - a future for Alzheimer's disease therapy? *Acta Biochim Pol.* 2004;51:431–444.
14. Huang Y, Skwarek-Maruszewska A, Horre K, et al. Loss of GPR3 reduces the amyloid plaque burden and improves memory in Alzheimer's disease mouse models. *Sci Transl Med.* 2015;7:309ra164.
15. Liao F, Hori Y, Hudry E, et al. Anti-ApoE antibody given after plaque onset decreases Aβ accumulation and improves brain function in a mouse model of Aβ amyloidosis. *J Neurosci.* 2014;34:7281–7292.
16. Murphy SR, Chang CC, Dogbevia G, et al. Acat1 knockdown gene therapy decreases amyloid-b in a mouse model of Alzheimer's disease. *Mol Ther.* 2013;21:1497–1506.
17. Kemppainen S, Lindholm P, Galli E, et al. Cerebral dopamine neurotrophic factor improves long-term memory in

APP/PS1 transgenic mice modeling Alzheimer's disease as well as in wild-type mice. *Behav Brain Res.* 2015;291:1–11.

18. Pascual-Lucas M, Viana da Silva S, Di Scala M, et al. Insulin-like growth factor 2 reverses memory and synaptic deficits in APP transgenic mice. *EMBO Mol Med.* 2014;6:1246–1262.

19. Nagahara AH, Mateling M, Kovacs I, et al. Early BDNF treatment ameliorates cell loss in the entorhinal cortex of APP transgenic mice. *J Neurosci.* 2013;33:15596–15602.

20. Tuszynski MH, Yang JH, Barba D, et al. Nerve growth factor gene therapy: activation of neuronal responses in Alzheimer disease. *JAMA Neurol.* 2015;72:1139–1147.

21. Wu Z, Asokan A, Samulski R. Adeno-associated virus serotypes: vector toolkit for human gene therapy. *Mol Ther.* 2006;14:316–327.

22. Fan DS, Ogawa M, Fujimoto KI, et al. Behavioral recovery in 6-hydroxydopamine-lesioned rats by cotransduction of striatum with tyrosine hydroxylase and aromatic L-amino acid decarboxylase genes using two separate adeno-associated virus vectors. *Hum Gene Ther.* 1998;9:2527–2535.

23. Chao H, Liu Y, Rabinowitz J, Li C, Samulski RJ, Walsh CE. Several log increase in therapeutic transgene delivery by distinct adeno-associated viral serotype vectors. *Mol Ther.* 2000;2:619–623.

24. Gao GP, Alvira MR, Wang L, Calcedo R, Johnston J, Wilson JM. Novel adeno-associated viruses from rhesus monkeys as vectors for human gene therapy. *Proc Natl Acad Sci USA.* 2002;99:11854–11859.

25. Mays LE, Vandenberghe LH, Xiao R, et al. Adeno-associated virus capsid structure drives CD4-dependent CD8+ T cell response to vector encoded proteins. *J Immunol.* 2009;182:6051–6060.

26. Mouri A, Noda Y, Hara H, Mizoguchi H, Tabira T, Nabeshima T. Oral vaccination with a viral vector containing Abeta cDNA attenuates age-related Abeta accumulation and memory deficits without causing inflammation in a mouse Alzheimer model. *FASEB J.* 2007;21:2135–2148.

27. Hara H, Monsonego A, Yuasa K, Adachi K, Xiao X, Takeda S. Development of a safe oral Abeta vaccine using recombinant adeno-associated virus vector for Alzheimer's disease. *J Alzheim Dis.* 2004;6:483–488.

28. Merkel SF, Andrews AM, Lutton EM, et al. Trafficking of AAV vectors across a model of the blood-brain barrier; a comparative study of transcytosis and transduction using primary human brain endothelial cells. *J Neurochem.* 2017;140:216–230.

29. Foust KD, Nurre E, Montgomery CL, Hernandez A, Chan CM, Kaspar BK. Intravascular AAV9 preferentially targets neonatal neurons and adult astrocytes. *Nat Biotechnol.* 2009;27:59–65.

30. Duque S, Joussemet B, Riviere C, et al. Intravenous administration of self-complementary AAV9 enables transgene delivery to adult motor neurons. *Mol Ther.* 2009;17:1187–1196.

31. Bevan AK, Duque S, Foust KD, Morales PR, Braun L, Schmelzer L, et al. Systemic gene delivery in large species for targeting spinal cord, brain, and peripheral tissues for pediatric disorders. *Mol Ther.* 2011;19:1971–1980.

32. Gray SJ, Matagne V, Bachaboina L, Yadav S, Ojeda SR, Samulski RJ. Preclinical differences of intravascular AAV9 delivery to neurons and glia: a comparative study of adult mice and nonhuman primates. *Mol Ther.* 2011;19:1058–1069.

33. Samaranch L, Salegio EA, San Sebastian W, et al. Adeno-associated virus serotype 9 transduction in the central nervous system of nonhuman primates. *Hum Gene Ther.* 2012;23:382–389.

34. Yang B, Li S, Wang H, et al. Global CNS transduction of adult mice by intravenously delivered rAAVrh.8 and rAAVrh.10 and nonhuman primates by rAAVrh.10. *Mol Ther.* 2014;22:1299–1309.

35. Xiao PJ, Samulski RJ. Cytoplasmic trafficking, endosomal escape, and perinuclear accumulation of adeno-associated virus type 2 particles are facilitated by microtubule network. *J Virol.* 2012;86:10462–10473.

36. Xiao PJ, Li C, Neumann A, Samulski RJ. Quantitative 3D tracing of gene-delivery viral vectors in human cells and animal tissues. *Mol Ther.* 2012;20:317–328.

37. Nicolson SC, Samulski RJ. Recombinant adeno-associated virus utilizes host cell nuclear import machinery to enter the nucleus. *J Virol.* 2014;88:4132–4144.

38. Kim HD, Maxwell JA, Kong FK, Tang DC, Fukuchi K. Induction of anti-inflammatory immune response by an adenovirus vector encoding 11 tandem repeats of Abeta1-6: toward safer and effective vaccines against Alzheimer's disease. *Biochem Biophys Res Commun.* 2005;336:84–92.

39. Zou J, Yao Z, Zhang G, et al. Vaccination of Alzheimer's model mice with adenovirus vector containing quadrivalent foldable Abeta(1–15) reduces Abeta burden and behavioral impairment without Abeta-specific T cell response. *J Neurol Sci.* 2008;272:87–98.

40. Li Y, Ma Y, Zong L-X, Xing X-N, Sha S, Cao Y-P. Intranasal inoculation with an adenovirus vaccine encoding ten repeats of Aβ 3–10 induces Th2 immune response against amyloid-β in wild-type mouse. *Neurosci Lett.* 2011;505:128–133.

41. Fang J, Qian JJ, Yi S, et al. Stable antibody expression at therapeutic levels using the 2A peptide. *Nat Biotechnol.* 2005;23:584–590.

42. Fang J, Yi S, Simmons A, et al. An antibody delivery system for regulated expression of therapeutic levels of monoclonal antibodies *in vivo. Mol Ther.* 2007;15:1153–1159.

43. Wuertzer CA, Sullivan MA, Qiu X, Federoff HJ. CNS delivery of vectored prion-specific single-chain antibodies delays disease onset. *Mol Ther.* 2008;16:481–486.

44. Wang YJ, Gao CY, Yang M, et al. Intramuscular delivery of a single chain antibody gene prevents brain Aβ deposition and cognitive impairment in a mouse model of Alzheimer's disease. *Brain Behav Immun.* 2010;24:1281–1293.

45. Kou J, Kim H, Pattanayak A, et al. Anti-amyloid-β single-chain antibody brain delivery via AAV reduces amyloid load but may increase cerebral hemorrhages in an Alzheimer's disease mouse model. *J Alzheimers Dis.* 2011;27:23–38.

46. Patel P, Kriz J, Gravel M, et al. Adeno-associated virus-mediated delivery of a recombinant single-chain antibody against misfolded superoxide dismutase for treatment of amyotrophic lateral sclerosis. *Mol Ther.* 2014;22:498–510.

47. DeMattos RB, Bales KR, Cummins DJ, Dodart JC, Paul SM, Holtzman DM. Peripheral anti-A beta antibody alters CNS and plasma A beta clearance and decreases brain A beta burden in a mouse model of Alzheimer's disease. *Proc Natl Acad Sci USA.* 2001;98:8850–8855.

48. Levites Y, Jansen K, Smithson LA, et al. Intracranial adeno-associated virus-mediated delivery of anti-pan amyloid beta, amyloid beta40, and amyloid beta42 single-chain variable fragments attenuates plaque pathology in amyloid precursor protein mice. *J Neurosci.* 2006;26:11923–11928.

49. Balazs AB, Chen J, Hong CM, Rao DS, Yang L, Baltimore D. Antibody-based protection against HIV infection by vectored immunoprophylaxis. *Nature.* 2012;481:81–84.

50. Ryan DA, Mastrangelo MA, Narrow WC, Sullivan MA, Federoff HJ, Bowers WJ. Abeta-directed single-chain antibody delivery via a serotype-1 AAV vector improves learning behavior and pathology in Alzheimer's disease mice. *Mol Ther.* 2010;18:1471–1481.

51. Shimada M, Abe S, Takahashi T, et al. Prophylaxis and treatment of Alzheimer's disease by delivery of an adeno-associated virus encoding a monoclonal antibody targeting the amyloid beta protein. *PLoS One.* 2013;8:e57606.

52. Wang YJ, Pollard A, Zhong JH, et al. Intramuscular delivery of a single chain antibody gene reduces brain Abeta burden in a mouse model of Alzheimer's disease. *Neurobiol Aging.* 2009;30:364–376.

53. Chai X, Wu S, Murray TK, Kinley R, Cella CV, Sims H. Passive immunization with anti-tau antibodies in two transgenic models: reduction of tau pathology and delay of disease progression. *J Biol Chem.* 2011;286:34457–34467.

54. Liu W, Zhao L, Blackman B, et al. Vectored intracerebral immunization with the anti-tau monoclonal antibody PHF1 markedly reduces tau pathology in mutant tau transgenic mice. *J Neurosci.* 2016;36:12425–12435.

55. Sondhi D, Johnson L, Purpura K, Monette S, Souweidane MM, Kaplitt MG. Long-term expression and safety of administration of AVrh.10hCLN2 to the brain of rats and nonhuman primates for the treatment of late infantile neuronal ceroid lipofuscinosis. *Hum Gene Ther Methods.* 2012;324:335.

56. Kells AP, Hadaczek P, Yin D, et al. Efficient gene therapy-based method for the delivery of therapeutics to primate cortex. *Proc Natl Acad Sci U S A.* 2009;106:2407–2411.

57. Baek RC, Broekman ML, Leroy SG, et al. AAV-mediated gene delivery in adult GM1-gangliosidase mice corrects lysosomal storage in CNS and improves survival. *PLoS One.* 2010;5:e13468.

58. Zhao L, Gottesdiener AJ, Parmar M, et al. Intracerebral adeno-associated virus gene delivery of APOE2 markedly reduces brain amyloid pathology in Alzheimer's disease mouse models. *Neurobiol Aging.* 2016;44:159–172.

59. Yoon SY, Bagel JH, O'Donnell PA, Vite CH, Wolfe JH. Clinical Improvement of alpha-mannosidosis cat following a single cisterna magna infusion of AAV. *Mol Ther.* 2016;24:26–33.

60. Zincarelli C, Soltys S, Rengo G, Rabinowitz JE. Analysis of AAV serotypes 1–9 mediated gene expression and tropism in mice after systemic injection. *Mol Ther.* 2008;16:1073–1080.

60a. Deverman BE, Pravdo PL, Simpson BP, et al. Cre-dependent selection yields AAV variants for widespread gene transfer to the adult brain. *Nat Biotechnol.* 2016;34:204–209.

61. Baker M, MacKenzie IR, Pickering-Brown SM, et al. Mutations in progranulin cause tau-negative frontotemporal dementia linked to chromosome 17. *Nature.* 2006;442:916–919.

62. Ahmed Z, MacKenzie IR, Hutton M, Dickson D. Progranulin in frontotemporal lobar degeneration and neuroinflammation. *J Neuroinflammation.* 2007;4:7.

63. Van Kampen JM, Kay DG. Progranulin gene delivery reduces plaque burden and synaptic atrophy in a mouse model of Alzheimer's disease. *PLoS One.* 2017;12:e0182896.

64. Mogi M, Harada N, Narabayashi H, Inagaki H, Minami M, Nagatsu T. Interleukin (IL)-1 beta, IL-2, IL-4, IL-6 and transforming growth factor-alpha levels are elevated in ventricular cerebrospinal fluid in juvenile parkinsonism and Parkinson's disease. *Neurosci Lett.* 1996;211:13–16.

65. Brouwers N, Sleegers K, Engelborghs S, et al. Genetic variability in progranulin contributes to risk for clinically diagnosed Alzheimer disease. *Neurology.* 2008;71:656–664.

66. Revuelta GJ, Rosso A, Lippa CF. Association between progranulin and beta-amyloid in dementia with Lewy bodies. *Am J Alzheimers Dis Other Demen.* 2008;23:488–493.

67. Martens LH, Zhang J, Barmada SJ, et al. Progranulin deficiency promotes neuroinflammation and neuron loss following toxin-induced injury. *J Clin Invest.* 2012;122:3955–3959.

68. Minami SS, Min SW, Krabbe G, et al. Progranulin protects against amyloid β deposition and toxicity in Alzheimer's disease mouse models. *Nat Med.* 2014;20:1157–1164.

69. Jin FY, Nathan C, Radzioch D, Ding A. Secretory leukocyte protease inhibitor: a macrophage product induced by and antagonistic to bacterial lipopolysaccharide. *Cell.* 1997;88:417–426.

70. Zhu JC, Nathan C, Jin W, et al. Conversion of proepithelin to epithelins: role of SLPI and elastase in host defense and wound repair. *Cell.* 2002;111:867–878.

71. Tang W, Lu Y, Tian QY, et al. The growth factor progranulin binds to TNF receptors and is therapeutic against inflammatory arthritis in mice. *Science.* 2011;332:478–484.

72. Shirotani K, Tsubuki S, Iwata N. Neprilysin degrades both amyloid û peptides 1±40 and 1±42 most rapidly and efficiently among thiorphan- and phosphoramidon-sensitive endopeptidases. *J Biol Chem.* 2001;276:21895–21901.

73. Iwata N, Tsubuki S, Takaki Y, et al. Metabolic regulation of brain Abeta by neprilysin. *Science*. 2001;292:1550–1552.

74. Spencer B, Marr RA, Rockenstein E, et al. Long-term neprilysin gene transfer is associated with reduced levels of intracellular Abeta and behavioral improvement in APP transgenic mice. *BMC Neurosci*. 2008;9:109.

75. Abdulhaqq SA, Weiner DB. DNA vaccines: developing new strategies to enhance immune responses. *Immunol Res*. 2008;42:219–232.

76. Ura T, Okuda K, Shimada M. Developments in viral vector-based vaccines. *Vaccines (Cold Spring Harbor)*. 2014;2:624–641.

77. Choi Y, Chang J. Viral vectors for vaccine applications. *Clin Exp Vaccine Res*. 2013;2:97–105.

78. Lemiale F, Kong WP, Akyurek LM, Ling X, Huang Y, Chakrabarti BK. Enhanced mucosal immunoglobulin A response of intranasal adenoviral vector human immunodeficiency virus vaccine and localization in the central nervous system. *J Virol*. 2003;77:10078–10087.

CHAPTER 8

Mitochondria-Targeted Nanoparticles: A Milestone or a Mirage in the Treatment of Alzheimer's Disease

ABSTRACT

Alzheimer's disease is the most common neurodegenerative disorder in which mitochondrial dysfunction is implicated. Mitochondria being the platform for vital metabolic activities play an essential role in brain function. However, its dysfunction is implicated in manifold diseases such as neurodegenerative diseases. This chapter highlights the importance of mitochondrial therapeutics, nanoparticles targeted at mitochondria, and their merits and demerits.

KEYWORDS

Alzheimer's disease; Intranasal delivery; Mitochondrial dysfunction; Nanoparticles.

INTRODUCTION

Mitochondria play a meticulous role in multifarious cellular activities such as energy metabolism, apoptosis, and steroid hormone synthesis.[1–3] Abnormally swollen, watery, and irregularly spread mitochondria in the cytoplasm of Alzheimer's disease (AD) patient's brain cells was reported in 1963.[4,5] In addition, mitochondrial DNA mutations followed by discrepancies and abnormalities in electron transport chain and other proteins were also found to contribute for approximately 63% of AD patients.[5,6]

AD is the most common neurodegenerative disease mostly found in women compared to men.[3] This may be due to the differences in mitochondrial function and neuronal energy metabolites.[3] In addition, augmented levels of antioxidants and increased life expectancy of females render them vulnerable to AD.[3,7–14] Recent reports have shown that among the elderly citizens of 71 years of age, 11% of men and 17% of women suffer from AD in the United States.[3,15] Similarly, in Europe, AD was found to be 3.31% in men and 7.13% in women.[3,16]

Mitochondrial dysfunction and impaired calcium homeostasis has long been implicated in the etiology of AD.[17–21] Additionally, Aβ was found to adversely affect mitochondrial respiration, and mitochondrial dysfunction also occurred earlier than Aβ extracellular accumulation.[18,22,23] Aβ was also found to provoke mitochondrial damage through enhanced production of ROS like superoxide anion, hydroxyl radicals, and hydrogen peroxide.[24,25,26] ROS induced mitochondrial damage being an earlier event than Aβ accumulation; copious studies targeted at curtailing oxidative stress played a pivotal role in AD therapeutics.[25,27,28] Accordingly, a few mitochondria-targeted antioxidants like MitoQ and SS31 were found to combat AD-associated oxidative stress and synaptic dysfunctions.[25,29–31]

Nanoparticles

A few biological riddles involved in mitochondrial targeting include biological barriers like intracellular spanning into mitochondria, outer and inner mitochondrial membranes, and mitochondrial toxicity.[5] A few nanosystems are targeted at mitochondria to overcome mitochondrial diseases, whereas a few other nanosystems are designed to mitigate mitochondrial toxicity.[5] Development of multifarious delivery systems which can deliver therapeutics to organelles by passing through BBB has been in progress and is still its infancy.[32–34] Although a few therapeutics have been discovered to treat mitochondrial dysfunction, yet their appropriate targeting to the mitochondria of the cell is a herculean task until now.[35,36]

Flurbiprofen Nanoparticles

Wealth of studies has shown that R-Flurbiprofen exhibits exemplary neuroprotection against AD by attenuating mitochondrial calcium overload instigated by Aβ toxicity.[18] Despite enhanced neuroprotection, inadequate brain entry after oral administration hindered its success.[18] With a view to overcoming this

Alzheimer's Disease Theranostics. https://doi.org/10.1016/B978-0-12-816412-9.00008-2
Copyright © 2019 Elsevier Inc. All rights reserved.

biological riddle, serum albumin nanoparticles (NPs) were designed which enhanced brain delivery via nose to brain route.[18] Aβ42 oligomers provoke enormous influx of calcium ions into neuronal cells eventually resulting in cell death.[18,37] A few NSAIDs like R-Flurbi-profen which mimic mitochondrial uncouplers attenuate calcium uptake at lower concentrations due to the presence of ionizable carboxylic group.[18,38]

Nattokinase Nanoparticles

Nattokinase, a protein with a plethora of therapeutic benefits such as antiamyloid activity, antifibrinolytic activity, and antithrombotic activity, is a robust AD therapeutic. A few studies reported the amyloid degrading efficacy of nattokinase *in vitro*.[39–41] However, its low stability impeded its success. With a view to overcoming this biological riddle, a biocompatible poly(lactic) glycolide NPs were designed in which nattokinase was enveloped. Surprisingly, no adverse effects were observed in enzyme activity after encapsulation.[41] Furthermore, adequate amelioration in its stability via the formation of a shell around it in an organic solvent and pharmacokinetic properties were accomplished by encapsulation in NPs.[41–43] PLGA has been an established polymer due to its appreciable biocompatibility and biodegradability.[41] It is metabolized in the body through the Kreb's cycle and broken down into glycolic acid and lactic acid.[41,44] Therefore, it was approved by the United States Food and Drug Administration (USFDA) and European Medicines Agency for its use in development of NPs.[41,45] PLGA nattokinase NPs are conjugated to Tet1 peptide of 12 amino acids (HLNILSTLWKYR) which shows enhanced affinity toward neuronal cells. Therefore, they feasibly interact with motor neurons and target Aβ.[41]

Intranasal Delivery

Growing lines of evidence has shown that intranasal administration can localize multifarious peptides in the brain within a short time period.[18,55] Additionally, it was observed that 125I labeled albumin was localized in all parts of the brain within 5 min after intranasal administration.[18,46] The probable underlying mechanism of albumin passage into brain is its spanning via intercellular gaps in the olfactory epithelium and reaching brain directly.[18,47] Albumin with its antioxidative property showed enhanced respiration in isolated rodent brain mitochondria.[18,48] Furthermore, biosafety and compatibility made albumin a potential carrier for brain-targeted therapeutics.[18,49]

Although substantial therapeutic efficacy of R-Flurbiprofen was observed in patients with mild AD in phase II clinical study, yet the robust efficacy was not found in phase III clinical trials.[18,50–52] Remarkably low brain localization was found to be the major issue for such clinical failure.[18,53] In light of this, noninvasive nose-to-brain delivery was opted to avert blood-brain barrier and localize therapeutic agents in the brain successfully.[18,54–57]

Triphenylphosphonium Functionalization

Multifarious functionalized delivery systems such as polymeric NPs, metallic NPs, and liposomes are more effective compared to nonfunctionalized nanoparticulate platforms in targeting mitochondria.[5] In a study, superoxide dismutase (SOD)–loaded PLGA NPs showed enhanced therapeutic efficacy in human neuronal cells compared to SOD alone.[5,58] Brenza et al.,[34] developed apocynins (4-hydroxy-3-methoxy-acetophenone) with various carbon chain lengths and functionalized it with triphenylphosphonium (TPP) cation (Mito-Apo) and accomplished substantial mitochondrial localization of NP and significant protective efficacy.[34] NPs also exhibit internalization through caveolin and clathrin-mediated endocytosis.[34] Indeed, overall negative charge of NPs may facilitate the internalization through lipid rafts.[34,59,60] Studies have shown that mitochondria-targeted antioxidant apocynin comprising biodegradable polyanhydride NPs conferred significant protection against oxidative cell damage in a dopaminergic neuronal cell line, mouse primary cortical neurons, and a human mesencephalic cell line.[34]

PLGA-b-PEG-TPP-QD (poly(D,L, lactic-co-glycolic acid)-block-poly(ethylene glycol)-triphenylphosphonium-quantum dot) targeted at mitochondrial delivery of curcumin showed remarkably improved therapeutic efficacy compared to curcumin alone.[36,61,62] Despite successful delivery of therapeutics to mitochondria by a few metal oxide- and liposomal-based nanocarriers, scant information about their optimization and lack of FDA-approved biodegradable PLGA raise a few concerns in current research.[36,63–71] PEG improves systemic circulation, and TPP facilitates internalization of therapeutic in mitochondria.[36,72,73]

While a few delivery systems were targeted at mitochondria to overcome disease symptoms, a few delivery systems targeted to other organelles exerted toxicity in mitochondria. In line with this, single-walled carbon nanotubes (SWCNTs) targeted at lysosomes exerted mitochondrial toxicity at higher doses despite their profound therapeutic effect at lower doses against AD.[74] Interestingly, lysosomal or mitochondrial distribution of SWCNTs can be regulated by regulating the autophagy regulators.[18]

Cerium Oxide Nanoparticles

Furthermore, TPP-linked ceria (CeO_2) NPs which act as robust ROS scavengers in mitochondria by transitioning between Ce^{3+} and Ce^{4+} oxidation states and eventually attenuate cell death in 5XFAD AD model mice.[25] These NPs curtail reactive gliosis and mitochondrial dysfunction in mice.[25] These NPs exhibited stable hydrodynamic diameter and colloidal stability when incubated with phosphate-buffered saline (PBS) and Dulbeccos modified Eagle's Medium (DMEM) and 10% fetal bovine serum (FBS).[25]

In another study, Ce^{3+} and Ce^{4+} transition property has been exploited to overcome mitochondrial fission. Accordingly, nanoceria were employed which were profoundly localized in mitochondrial outer membrane and combated Aβ and peroxynitrite-instigated mitochondrial fission.[75] Peroxynitrite usually plays a pivotal role in Aβ aggregation and neurofibrillary tangle (NFT) formation.[75–78] This effect of mitochondrial fission and cell death has been induced through activation of dynamin-related-protein1 (DRP1), a huge GTPase that mediates mitochondrial fission by hitherto unknown molecular mechanisms.[75] However, DRP1 serine 616 phosphorylation plays a probable role. Cerium oxide NPs effectively eliminates superoxide anions, hydrogen peroxide, and peroxynitrite.[75] They also profoundly curtail ROS and reactive nitrogen species (RNS).[75]

CONCLUSION

Despite a dearth of studies on the molecular basis of mitochondrial dysfunction, a plethora of prior studies unraveled that interaction of Aβ with a few mitochondrial proteins like cyclophilin D, alcohol dehydrogenase, and ATP synthase contribute significantly to ROS generation.[25,79–81]

Although SWCNTs hold considerable promise for AD therapeutics, yet numerous studies showed the cytotoxicity of them.[74,82–94]

REFERENCES

1. Kuklinski B. *Mitochondrien: Symptome, Diagnose and Therapie*. 1st ed. Aurum; 2015.
2. Bakthavachalam P, Shanmugam PST. Mitochondrial dysfunction—silent killer in cerebral ischemia. *J Neurol Sci.* 2017;375:417–423.
3. Silaidos C, Pilatus U, Grewal R, et al. Sex-associated differences in mitochondrial function in human peripheral blood mononuclear cells (PBMCs) and brain. *Biol Sex Differ.* 2018;9:34.
4. Terry RD. The Fine structure of Neurofibrillary Tangles in Alzheimer's disease. *J Neuropathol Exp Neurol.* 1963;22:629–642.
5. Durazo SA, Kompella UB. Functionalized nanosystems for targeted mitochondrial delivery. *Mitochondrion.* 2012;12:190–201.
6. Coskun PE, Beal MF, Wallace DC. Alzheimer's brains harbor somatic mtDNA control-region mutations that suppress mitochondrial transcription and replication. *Proc Natl Acad Sci USA.* 2004;101:10726–10731.
7. Bachman DL, Wolf PA, Linn R, Knoefel JE, Cobb J, Belanger A. Prevalence of dementia and probable senile dementia of the Alzheimer type in the Framingham study. *Neurol.* 1992;42:115–119.
8. Vina J, Borras C. Women live longer than men: understanding molecular mechanisms offers opportunities to intervene by using estrogenic compounds. *Antioxid Redox Signal.* 2010;13:269–278.
9. Mandal PK, Tripathi M, Sugunan S. Brain oxidative stress: detection and mapping of anti-oxidant marker 'glutathione' in different brain regions of healthy male/female, MCI and Alzheimer patients using non-invasive magnetic resonance spectroscopy. *Biochem Biophys Res Commun.* 2012;417:43–48.
10. Gaignard P, Savouroux S, Liere P, et al. Effect of sex differences on brain mitochondrial function and its suppression by ovariectomy and in aged mice [eng]. *Endocrinology.* 2015;156:2893–2904.
11. Demarest TG, McCarthy MM. Sex differences in mitochondrial (dys)function: implications for neuroprotection. *J Bioenerg Biomembr.* 2015;47:173–188.
12. Gabelli C, Cademo A. Gender differences in cognitive decline and Alzheimer's disease. *Ital J Gender-Specific Med.* 2015;1:21–28.
13. Ostan R, Monti D, Gueresi P, Bussolotto M, Franceschi C, Baggio G. Gender, aging and longevity in humans: an update of an intriguing/neglected scenario paving the way to a gender-specific medicine [eng]. *Clin Sci (Lond).* 2016;130:1711–1725.
14. Austad SN, Fischer KE. Sex differences in lifespan [eng]. *Cell Metabol.* 2016;23:1022–1033.
15. Wortmann M. World Alzheimer report 2014: dementia and risk reduction. *Alzheimers Dement.* 2015;11:P837.
16. Niu H, Alvarez-Alvarez I, Guillen-Grima F, Aguinaga-Ontoso I. Prevalence and incidence of Alzheimer's disease in Europe: a meta-analysis. *Neurol.* 2017;32:523–532.
17. LaFerla FM. Calcium dyshomeostasis and intracellular signalling in Alzheimer's disease. *Nat Rev Neurosci.* 2002;3:862–872.
18. Wong RL, Ho PC. Role of serum albumin as a nanoparticulate carrier for nose-to-brain delivery of R-flurbiprofen: implications for the treatment of Alzheimer's disease. *J Pharm Pharmacol.* 2018;70:59–69.
19. Celsi F, Pizzo P, Brini M, et al. Mitochondria, calcium and cell death: a deadly triad in neurodegeneration. *Biochim Biophys Acta.* 2009;1787:335–344.

20. Supnet C, Bezprozvanny I. Neuronal calcium signaling, mitochondrial dysfunction, and Alzheimer's disease. *J Alzheimers Dis.* 2010;20:S487–S498.

21. Hung CH, Ho YS, Chang RC. Modulation of mitochondrial calcium as a pharmacological target for Alzheimer's disease. *Ageing Res Rev.* 2010;9:447–456.

22. Casley CS, Canevari L, Land JM, Clark JB, Sharpe MA. Beta-amyloid inhibits integrated mitochondrial respiration and key enzyme activities. *J Neurochem.* 2002;80: 91–100.

23. Chang KL, Pee HN, Tan WP, et al. Metabolic profiling of CHO-AbPP695 cells revealed mitochondrial dysfunction prior to amyloid-b pathology and potential therapeutic effects of both PPARc and PPARa agonisms for Alzheimer's disease. *J Alzheimers Dis.* 2015;44:215–231.

24. Westermann B. Nitric oxide links mitochondrial fission to Alzheimer's disease. *Sci Signal.* 2009;2.

25. Kwon HJ, Cha MY, Kim D, et al. Mitochondria-targeting ceria nanoparticles as antioxidants for Alzheimer's disease. *ACS Nano.* 2016;10:2860–2870.

26. Bartley MG, Marquardt K, Kirchhof D, Wilkins HM, Patterson D, Linseman DA. Overexpression of amyloid-β protein precursor induces mitochondrial oxidative stress and activates the intrinsic apoptotic cascade. *J Alzheimers Dis.* 2012;28:855–868.

27. Shaw LM, Vanderstichele H, Knapik-Czajka M, et al. Cerebrospinal fluid biomarker signature in Alzheimer's disease neuroimaging initiative subjects. *Ann Neurol.* 2009;65:403–413.

28. Du H, Guo L, Yan S, Sosunov AA, McKhann GM, Yan SS. Early deficits in synaptic mitochondria in an Alzheimer's disease mouse model. *Proc Natl Acad Sci USA.* 2010;107: 18670–18675.

29. Dickinson BC, Chang CJ. A targetable fluorescent probe for imaging hydrogen peroxide in the mitochondria of living cells. *J Am Chem Soc.* 2008;130:9638–9639.

30. Eckert A, Schulz KL, Rhein V, Gootz J. Convergence of amyloid-β and tau pathologies on mitochondria *in vivo*. *Mol Neurobiol.* 2010;41:107–114.

31. McManus MJ, Murphy MP, Franklin JL. The mitochondria-targeted antioxidant MitoQ prevents loss of spatial memory retention and early neuropathology in a transgenic mouse model of Alzheimer's disease. *J Neurosci.* 2011;31:15703–15715.

32. Ross KA, Brenza TM, Binnebose AM, et al. Nano-enabled delivery of diverse payloads across complex biological barriers. *J Contr Release.* 2015;219:548–559.

33. Mallapragada SK, Brenza TM, McMillan JM, et al. Enabling nanomaterial, nanofabrication and cellular technologies for nanoneuro medicines. *Nanomedicine.* 2015;11: 715–729.

34. Brenza TM, Ghaisas S, Ramirez JEV, et al. Neuronal protection against oxidative insult by polyanhydride nanoparticle-based mitochondria-targeted antioxidant therapy. *Nanomedicine.* 2017;13:809–820.

35. Fulda S, Galluzzi L, Kroemer G. Targeting mitochondria for cancer therapy. *Nat Rev Drug Discov.* 2010;9:447–464.

36. Marrache S, Dhar S. Engineering of blended nanoparticle platform for delivery of mitochondria-acting therapeutics. *Proc Natl Acad Sci Unit States Am.* 2012;109:16288–16293.

37. Sanz-Blasco S, Valero RA, Rodriguez-Crespo I, Villalobos C, Nunez L, et al. Mitochondrial Ca^{2+} overload underlies Abeta oligomers neurotoxicity providing an unexpected mechanism of neuroprotection by NSAIDs. *PloS One.* 2008;3:e2718.

38. Calvo-Rodriguez M, Nunez L, Villalobos C, et al. Nonsteroidal anti-inflammatory drugs (NSAIDs) and neuroprotection in the elderly: a view from the mitochondria. *Neural Regen Res.* 2015;10:1371–1372.

39. Hsu RL, Lee KT, Wang JH, Lee LY, Chen RP. Amyloid-degrading ability of nattokinase from *Bacillus subtilis* natto. *J Agric Food Chem.* 2009;57:503–508.

40. Wang C, Du M, Zheng D, Kong F, Zu G, Feng Y. Purification and characterization of nattokinase from *Bacillus subtilis* natto B-12. *J Agric Food Chem.* 2009;57:9722–9729.

41. Bhatt PC, Verma A, Al-Abbasi FA, Anwar F, Kumar V, Panda BP. Development of surface-engineered PLGAnanoparticulate-delivery system of Tet1-conjugated nattokinase enzyme for inhibition of Aβ40 plaques in Alzheimer's disease. *Int J Nanomed.* 2017;12:8749–8768.

42. Manish M, Rahi A, Kaur M, Bhatnagar R, Singh SA. Single-dose PLGA encapsulated protective antigen domain 4 nanoformulation protects mice against *Bacillus anthracis* spore challenge. *PloS One.* 2013;8:e61885.

43. Manish M, Bhatnagar R, Singh S. Preparation and characterization of PLGA encapsulated protective antigen domain 4 nanoformulation. *Methods Mol Biol.* 2016;1404:669–681.

44. Idelchik GM, Varon J. Hypocarbia, therapeutic hypothermia, and mortality: the Kreb cycle is key. *Am J Emerg Med.* 2014;32:643–644.

45. Tao W, Zeng X, Liu T, et al. Docetaxel-loaded nanoparticles based on star-shaped mannitol-core PLGA-TPGS diblock copolymer for breast cancer therapy. *Acta Biomater.* 2013;9:8910–8920.

46. Falcone JA, Salameh TS, Yi X, et al. Intranasal administration as a route for drug delivery to the brain: evidence for a unique pathway for albumin. *J Pharmacol Exp Therapeut.* 2014;351:54–60.

47. Meredith ME, Salameh TS, Banks WA. Intranasal delivery of proteins and peptides in the treatment of neurodegenerative diseases. *AAPS J.* 2015;17:780–787.

48. Panov AV, Vavilin VA, Lyakhovich VV, Brooks BR, Bonkovsky HL. Effect of bovine serum albumin on mitochondrial respiration in the brain and liver of mice and rats. *Bull Exp Biol Med.* 2010;149:187–190.

49. Foote M. Using nanotechnology to improve the characteristics of antineoplastic drugs: improved characteristics of nab-paclitaxel compared with solvent-based paclitaxel. *Biotechnol Annu Rev.* 2007;13:345–357.

50. Wilcock GK, Black SE, Hendrix SB, et al. Efficacy and safety of tarenflurbil in mild to moderate Alzheimer's disease: a randomized phase II trial. *Lancet Neurol.* 2008;7: 483–493.

51. Wilcock GK, Black SE, Balch AA, et al. Safety and efficacy of tarenflurbil in subjects with mild Alzheimer's disease: results from an 18-month international multi-center phase 3 trial. *Alzheimers Dement.* 2009;5:P86.

52. Green RC, Schneider LS, Amato DA, et al. Effect of tarenflurbil on cognitive decline and activities of daily living in patients with mild Alzheimer disease: a randomized controlled trial. *J Am Med Assoc.* 2009;302:2557–2564.

53. Imbimbo BP. Why did tarenflurbil fail in Alzheimer's disease? *J Alzheimers Dis.* 2009;17:757–760.

54. Hanson LR, Frey WH. Intranasal delivery bypasses the blood-brain barrier to target therapeutic agents to the central nervous system and treat neurodegenerative disease. *BMC Neurosci.* 2008;9:S5.

55. Lochhead JJ, Thorne RG. Intranasal delivery of biologics to the central nervous system. *Adv Drug Deliv Rev.* 2012;64:614–628.

56. Pardeshi CV, Belgamwar VS. Direct nose to brain drug delivery via integrated nerve pathways bypassing the blood-brain barrier: an excellent platform for brain targeting. *Expert Opin Drug Deliv.* 2013;10:957–972.

57. Sood S, et al. Intranasal therapeutic strategies for management of Alzheimer's disease. *J Drug Target.* 2014;22: 279–294.

58. Reddy MK, Wu L, Kou W, Ghorpade A, Labhasetwar V. Superoxide dismutase-loaded PLGA nanoparticles protect cultured human neurons under oxidative stress. *Appl Biochem Biotechnol.* 2008;151:565–577.

59. Bannunah AM, Vllasaliu D, Lord J, Stolnik S. Mechanisms of nanoparticle internalization and transport across an intestinal epithelial cell model: effect of size and surface charge. *Mol Pharm.* 2014;11:4363–4373.

60. Phanse Y, Lueth P, Ramer-Tait AE, et al. Cellular internalization mechanisms of polyanhydride particles: implications for rational design of drug delivery vehicles. *J Biomed Nanotechnol.* 2016;12:1544–1552.

61. Mulik RS, Monkkonen J, Juvonen RO, Mahadik KR, Paradkar AR. ApoE3-mediated poly(butyl) cyanoacrylate nanoparticles containing curcumin: study of enhanced activity of curcumin against beta amyloid-induced cytotoxicity using *in vitro* cell culture model. *Mol Pharm.* 2010;7:815–825.

62. Ono K, Hasegawa K, Naiki H, Yamada M. Curcumin has potent anti-amyloidogenic effects for Alzheimer's beta-amyloid fibrils *in vitro*. *J Neurosci Res.* 2004;75:742–750.

63. Esumi K, Takei N, Yoshimura T. Antioxidant potentiality of gold-chitosan nanocomposites. *Colloids Surf B.* 2003;32:117–123.

64. Esumi K, Houdatsu H, Yoshimura T. Antioxidant action by gold-PAMAM dendrimer nanocomposites. *Langmuir.* 2004;20:2536–2538.

65. Paunesku T, Vogt S, Lai B, et al. Intracellular distribution of TiO2-DNA oligonucleotide nanoconjugates directed to nucleolus and mitochondria indicates sequence specificity. *Nano Lett.* 2007;7:596–601.

66. Kajita M, Hikosaka K, Iitsuka M, Kanayama A, Toshima N, Miyamoto Y. Platinum nanoparticle is a useful scavenger of superoxide anion and hydrogen peroxide. *Free Radic Res.* 2007;41:615–626.

67. Hikosaka K, Kim J, Kajita M, Kanayama A, Miyamoto Y. Platinum nanoparticles have an activity similar to mitochondrial NADH:ubiquinone oxidoreductase. *Colloids Surf B Biointerfaces.* 2008;66:195–200.

68. Boddapati SV, D'Souza GGM, Erdogan S, Torchilin VP, Weissig V. Organelle targeted nanocarriers: specific delivery of liposomal ceramide to mitochondria enhances its cytotoxicity *in vitro* and *in vivo*. *Nano Lett.* 2008;8:2559–2563.

69. Patel NR, Hatziantoniou S, Georgopoulos A, et al. Mitochondria-targeted liposomes improve the apoptotic and cytotoxic action of sclareol. *J Liposome Res.* 2010;20:244–249.

70. Wang LM, Liu Y, Li W, et al. Selective targeting of gold nanorods at the mitochondria of cancer cells: implications for cancer therapy. *Nano Lett.* 2011;11:772–780.

71. Sharma A, Soliman GM, Al-Hajaj N, Sharma R, Maysinger D, Kakkar A. Design and evaluation of multifunctional nanocarriers for selective delivery of coenzyme Q10 to mitochondria. *Biomacromolecules.* 2012;13:239–252.

72. Smith RA, Porteous CM, Gane AM, Murphy MP. Delivery of bioactive molecules to mitochondria *in vivo*. *Proc Natl Acad Sci USA.* 2003;100:5407–5412.

73. Beletsi A, Panagi Z, Avgoustakis K. Biodistribution properties of nanoparticles based on mixtures of PLGA with PLGA-PEG diblock copolymers. *Int J Pharm.* 2005;298:233–241.

74. Yang Z, Zhang Y, Yang Y, et al. Pharmacological and toxicological target organelles and safe use of single-walled carbon nanotubes as drug carriers in treating Alzheimer disease. *Nanomed Nanotechnol Biol Med.* 2010;6: 427–441.

75. Dowding JM, Song W, Bossy K, et al. Cerium oxide nanoparticles protect against Aβ-induced mitochondrial fragmentation and neuronal cell death. *Cell Death Differ.* 2014;21:1622–1632.

76. Knott AB, Bossy-Wetzel E. Nitric oxide in health and disease of the nervous system. *Antioxid Redox Signal.* 2009;11:541–554.

77. Guix FX, Ill-Raga G, Bravo R, et al. Amyloid-dependent triosephosphate isomerase nitrotyrosination induces glycation and tau fibrillation. *Brain.* 2009;132:1335–1345.

78. Smith MA, Richey Harris PL, Sayre LM, Beckman JS, Perry G. Widespread peroxynitrite-mediated damage in Alzheimer's disease. *J Neurosci.* 1997;17:2653–2657.

79. Lustbader JW, Cirilli M, Lin C, et al. ABAD directly links aß to mitochondrial toxicity in Alzheimer's disease. *Science.* 2004;304:448–452.

80. Du H, Guo L, Fang F, Chen D, Sosunov AA, McKhann GM. Cyclophilin D deficiency attenuates mitochondrial and neuronal perturbation and ameliorates learning and memory in Alzheimer's disease. *Nat Med.* 2008;14:1097–1105.

81. Schmidt C, Lepsverdize E, Chi S, et al. Amyloid precursor protein and amyloid-β-peptide bind to ATP synthase and regulate its activity at the surface of neural cells. *Mol Psychiatr.* 2008;13:953–969.

82. Shvedova AA, Castranova V, Kisin ER, et al. Exposure to carbon nanotube material: assessment of nanotube cyto-toxicity using human keratinocyte cells. *J Toxicol Environ Health*. 2003;66:1909–1926.

83. Shvedova AA, Kisin ER, Mercer R, et al. Unusual inflammatory and fibrogenic pulmonary responses to single-walled carbon nanotubes in mice. *Am J Physiol Lung Cell Mol Physiol*. 2005;289:L698–L708.

84. Lam CW, James JT, McCluskey R, Hunter RL. Pulmonary toxicity of single-wall carbon nanotubes in mice 7 and 90 days after intratracheal instillation. *Toxicol Sci*. 2004;77:126–134.

85. Cui D, Tian F, Ozkan CS, Wang M, Gao H. Effect of single wall carbon nanotubes on human HEK293 cells. *Toxicol Lett*. 2005;155:73–85.

86. Muller J, Huaux F, Moreau N, et al. Respiratory toxicity of multi-wall carbon nanotubes. *Toxicol Appl Pharmacol*. 2005;207:221–231.

87. Heller D, Baik S, Eurell T, Strano M. Single-walled carbon nanotube spectroscopy in live cells: towards long-term labels and optical sensors. *Adv Mater*. 2005;17:2793–2799.

88. Sato Y, Yokoyama A, Shibata K, et al. Influence of length on cytotoxicity of multi-walled carbon nanotubes against human acute monocytic leukemia cell line THP-1 *in vitro* and subcutaneous tissue of rats *in vivo*. *Mol Biosyst*. 2005;1:176–182.

89. Bottini M, Bruckner S, Nika K, et al. Multi-walled carbon nanotubes induce T lymphocyte apoptosis. *Toxicol Lett*. 2006;160:121–126.

90. Smart SK, Cassady AI, Lu GQ, Martin DJ. The biocompatibility of carbon nanotubes. *Carbon*. 2006;44:1034–1047.

91. Kagan VE, Tyurina YY, Tyurin VA, et al. Direct and indirect effects of single walled carbon nanotubes on RAW 264.7 macrophages: role of iron. *Toxicol Lett*. 2006;165:88–100.

92. Amatore C, Arbault S, Cristhina D, et al. Stimulates or attenuates reactive oxygen and nitrogen species (ROS, RNS) production depending on cell state: quantitative amperometric measurements of oxidative bursts at PLB-985 and RAW 264.7 cells at the single cell level. *J Electroanal Chem*. 2008;615:34–44.

93. Garza KM, Soto KF, Murr LE. Cytotoxicity and reactive oxygen species generation from aggregated carbon and carbonaceous nanoparticulate materials. *Int J Nanomed*. 2008;3:83–94.

94. Lacerda L, Soundararajan A, Pastorin G, et al. Dynamic imaging of functionalized multi-walled carbon nanotube systemic circulation and urinary excretion. *Adv Mater*. 2008;20:225–230.

CHAPTER 9

Nanoparticles: The Double-Edged Swords

ABSTRACT

Abundance of research accentuated the burgeoning need to develop robust theranostics for Alzheimer's disease (AD) due to the alarming surge in AD patients. Since the molecular underpinnings of AD are incompletely understood, the development of appropriate theranostics has been a mirage till date. Several therapeutic avenues available to treat AD are drugs, natural compounds, metal chelators, and growth factors. However, these therapeutic avenues repose challenges such as short systemic circulation, entry through the blood-brain barrier, and bioavailability. To overcome the above mentioned challenges, nanotechnology has extensively been used. Despite the discovery of dynamic biomaterials using nanotechnology to treat AD, lack of appropriate material has been an Achilles' heel till date. This chapter highlights the merits and demerits of currently available nanotechnological therapeutic arsenals such as drug delivery systems (DDSs) for drugs, biomolecules, metal chelators, and tissue engineering scaffolds.

KEYWORDS

Alzheimer's disease; Carbon nanotubes; Hydrogels; Polymeric nanoparticles; Tissue engineering scaffolds.

INTRODUCTION

Wealth of studies availed in Alzheimer's disease (AD) treatment include a plethora of drugs such as galantamine and rivastigmine,[1] and natural compounds such as curcumin,[2] resveratrol,[3] vitamin C,[4] and carotenoids.[5-7] Although they showed adequate therapeutic efficacy in manifold cells and animal and human models, yet limitations such as brain targeting, poor solubility, bioavailability, and toxicity issues impeded their success considerably.[8] To circumvent these biological riddles, multifarious biomaterials have been extensively used in the treatment of AD, which encompasses polymeric nanoparticles (NPs), liposomes, hydrogels, dendrimers, and solid lipid NPs.[9] NPs are condensed particles with the therapeutic either loaded in it or conjugated on the surface.[9-11] They are capable of releasing the therapeutic in a controlled manner for a few hours to days based on the nanoparticulate system.[9] However, a few hitherto unresolved challenges hinder their success significantly. One of these major challenges is formation of protein corona.[12] To overcome the protein corona and its interaction with plasma proteins of the patients, a meticulous colorimetric sensor was developed which succeeded in detecting AD and multiple sclerosis accurately.[13]

Nanotechnology

Limitations with the other delivery devices prompted the advent of nanodevices for therapeutic delivery to the brain primarily by employing biopolymer (polysaccharides) carriers with targeting ligands.[14] Nanotechnology which entails designing of materials with 1–1000 nm, profoundly enhances the therapeutic efficacy of loaded compounds in turn overcoming drug-associated toxicity and accomplishing targeted delivery.[14,15] Other attractive features include adjustable degradation rate, enhanced drug loading capacity, and blood-brain barrier (BBB) permeability.[14] A growing body of evidence also suggests that PEG substantially averts the enzymatic degradation of loaded drugs in NPs and also prevents its recognition from the immune system.[14,16] Vital transport mechanisms of NPs through BBB include opening of tight junctions and improving permeabilization,[17,18] transcytosis,[19] endocytosis,[20] and cooperation of two or more of these mechanisms.[18,21] Paracellular transport of a few drugs/NPs through BBB has been hindered by the tight junctions of BBB.[9] With a view to overcoming this biological riddle, a few permeability enhancers such as peptidase inhibitors, angiotensin II, bradykinins, and antimicrotubule agents were extensively employed.[9,22-25]

Polymeric Materials

Polymers that are profusely used to design nanomaterials include poly(lactide-co-glycolide) (PLGA), poly(lactic acid) (PLA), chitosan, gelatin, poly(caprolactone),

Alzheimer's Disease Theranostics. https://doi.org/10.1016/B978-0-12-816412-9.00009-4
Copyright © 2019 Elsevier Inc. All rights reserved.

poly(alkyl cyanoacrylates).[26] PLGA-functionalized quercetin NPs effectively perturbed Zn^{2+}-$A\beta_{42}$ interaction in both *in vitro* and *in vivo*, thus ameliorating AD symptoms.[27] Anthocyanin-loaded PLGA NPs ameliorated AD symptoms by hampering AD markers such as amyloid precursor protein (APP), β-site amyloid precursor protein cleaving enzyme 1 (BACE-1), and apoptotic proteins such as Bax, Bcl2, and Caspase-3.[28] Epigallocatechin-3-gallate (EGCG), a pivotal polyphenol abundant in green tea, was loaded in orally delivered nanolipidic particles to enhance its bioavailability by twofold and reverse AD symptoms due to its appreciable antioxidant and antiamyloidogenic properties.[26,29]

Similarly, growing lines of evidence showed that poly(butyl cyanoacrylate) delivered dalargin, loperamide, methotrexate, doxorubicin, and temozolomide to the brain successfully.[9,30–33] It has also been emphasized that apolipoproteins conjugated on the NPs' surface play a pivotal role in the BBB permeability.[9] The apolipoproteins adsorb on low density lipoprotein (LDL) receptors of microvessel endothelial cells and span BBB by receptor mediated transcytosis.[9,34] However, these NPs were found to cause a mild perturbance to the endothelial cells which can feasibly be recovered.[9,35] Approximately 3–20 fold enhanced delivery of dalargin and rhodamine was observed when used in combination with PBCA NPs.[9,30,36]

Conjugation of a surfactant polysorbate 80 to NPs potentiates them to span BBB, mimic LDL receptors, and enter the brain.[14,37] Yin et al.[38] designed sialic acid–modified selenium NPs coupled to peptide B6 (B6-SA-SeNPs) which successfully spanned BBB and attenuated Aβ fibrillation.[38] Multiple lines of evidence also showed that PBCA loaded with (125) I-clioquinol (CQ, 5-chloro-7-iodo-8-hydroxyquinoline), an Aβ affinity drug, effectively transported through the BBB and locate Aβ plaques both *in vitro* and *in vivo*.[39] Mounting evidence has shown that the chitosan and acetylated chitosan oligosaccharides showed appreciable neuroprotective efficacy against AD.[40] Surface functionalization of NPs achieved by adsorption or chemical grafting of polyethylene glycol (PEG), poloxamers or other molecules significantly enhance their circulation lifetime and increase drug delivery via BBB.[41] A biocompatible and bioresorbable liraglutide, a glucagon-like peptide analog–loaded PLA, and gelatin nanofibers succeeded in ameliorating neuroprotection via long-term release of liraglutide.[42]

Carbon Nanotubes
Carbon nanotubes (CNTs) designed recently showed substantial enhancement of engrafted neural stem cells' therapeutic efficacy against TMT-induced neurodegeneration in a rat model.[43] Recently, an antibody imitating acetylcholine-encapsulated CNTs associated with aniline were designed for the first time which can successfully be used to treat AD.[44] Ascorbic acid levels profoundly influence the pathology of brain diseases although the underlying mechanism is elusive. Therefore, a sensitive CNT fiber with ascorbic acid oxidation efficacy was developed recently that can evaluate ascorbic acid levels and diagnose AD.[45] Zhu et al.[46] developed an innovative CNT-based sandwich-type biosensor for the identification of α-1 antitrypsin (AAT, a known biomarker for AD) using 3, 4, 9, 10-perylene tetracarboxylic acid/carbon nanotubes (PTCA-CNTs) as sensing components and alkaline phosphatase labeled AAT antibody-coated silver NPs as a signal amplifier.

Hydrogels
Hydrogels being robust drug/therapeutic reservoirs serve as amenable tools to overcome several toxicity issues by sustained drug release.[47] They have been considered appealing in the treatment of AD due to their drug or cell encapsulation and release ability and noninvasive properties.[26,48] In line with this, development of three-dimensional (3D) *in vitro* neural lineage cultures by using hydrogels, solid porous polymers, and fibrous materials is of utmost important to study them as disease models such as AD.[49] Recently designed self-healing thermoreversible hydrogels using Fmoc-protected peptides were proven to be nontoxic and thixotropic and efficiently combat AD.[50] Injectable gel has long been extensively used as a controlled therapeutic releasing DDS for multifarious drugs and biomolecules that are to be spanned through BBB.[26,51–55] Polymer-based composites are of significant importance due to their merits in therapeutic encapsulation, tissue engineering, and targeted drug delivery.[26,56–58]

Tissue Engineering Scaffolds
The challenges associated with the development of aged AD brain *in vitro* are a replication of augmented levels of soluble and insoluble toxic Aβ comprising tissue.[59–62] To overcome this, a genetically engineered 3D cell culture was designed which showed AD pathological hallmarks such as Aβ accumulation and NFT.[60,63] Kim et al.[60] designed 3D neural culture system that replicated AD pathology such as development of Aβ accumulation in 1–2 days and tau proteins in 10–14 weeks more appropriately compared to *in vivo* models.[60] The culture model was carved by two vital components: human neural progenitor cells with remarkable ability

to generate Aβ species and Matrigel-based 3D culture system that facilitates Aβ aggregation. This technology being the first human cellular model may feasibly be used to study molecular underpinnings of AD such as Aβ aggregation and deposition and hyperphosphorylation of tau. Furthermore, it can also facilitate drug testing.[60] However, a few demerits such as lack of microglial cells in the system, inability to design specific brain regions, and inability to target early stage AD limit its success.[60]

Delivery Systems for Growth Factors

Multiple growth factors such as insulin-like growth factor (IGF), basic fibroblast growth factor (bFGF), and nerve growth factor (NGF) showed substantial amelioration of AD pathology in animal models.[14,64] Nevertheless, BBB, enzymatic degradation, and denaturation hinder their brain entry thus halting their success.[14,65] To overcome these issues, Herran et al.[66] designed vascular endothelial growth factor–loaded nanospheres (VEGF-NS) by double emulsion solvent evaporation technique using PLGA. These VEGF-NS were introduced into APP/Ps1 transgenic mice brain by craniotomy. Surprisingly, profound cell proliferation was observed in the hippocampus region, thus opening a new therapeutic maneuver to treat AD.[66]

Neurotrophic factors are neuroprotective in function, thus playing a vital role in reversing AD symptoms.[67] However, a few impediments such as plasma clearance within a short time period and enzymatic degradation were found to curb their success.[68,69] To overcome these issues, a few invasive and expensive strategies were employed such as convention-enhanced delivery and intrathecal and intraventricular injection. Since these procedures involve hospitalization and are inappropriate for several AD patients, injectable heparin-conjugated Tetronic PCL micelles were designed which showed approximately 70% of basic fibroblast growth factor (bFGF) loading efficiency, in turn achieving sustained release for 2 months *in vitro*.[67,70] Furthermore, nerve growth factor poly(ethylene glycol)-poly(lactic-co-glycolic acid) NPs and neural stem cell (NSC) transplantation showed remarkable amelioration of AD symptoms in rats.[71]

Delivery Systems for Chelators

Numerous metals implicated in AD include iron, copper, aluminum, and zinc. Metal chelation has long been extensively used to treat AD.[6,7] However, the toxicity exerted by chelators, lack of specificity, and short systemic circulation hamper their efficacy remarkably.[72] The advent of nanotechnology showed enhanced therapeutic efficacy of the available metal chelating drugs. The promising strategy developed till date to treat AD was brain targeting of Cu^{2+}/Zn^{2+} chelator, clioquinol to the wild-type mouse brain to solubilize Aβ.[9,73] Manifold ROS-generating metals were chelated by employing multifaceted NPs which showed lower toxicity and higher stability.[9,74,75]

A recently designed multifaceted nanoprobe with two counterparts such as upconversion NPs for the identification and upconversion luminescence imaging of copper when subjected to 980 nm and the chelator 8-hydroxyquinoline 2 carboxylic acid for copper chelation significantly ameliorated AD symptoms.[76] Similarly, recently designed iminodiacetic acid–conjugated NPs showed enhanced zinc chelation and significantly curtailed Aβ fibrillation in SH-SY5Y cells.[77] In another study, ligand-bound nanoliposomes surface functionalized by using copper acetate (CuAc), zinc acetate (ZnAc), ethylene diamine tetraacetic acid (EDTA), and histidine showed successful metal chelation and Aβ aggregation.[78]

DEMERITS OF BIOMATERIALS

NPs that are extensively used to overcome not only AD but also several other diseases often exert severe systemic toxicity.[26,79,80] In addition, the extremely expensive therapeutic strategies employed to treat brain diseases accounts to US$ 100 million to US$ 1 billion before reaching the market. However, only 3%–5% of the therapeutics reaches the market due to the BBB impediment.[18] Despite the adequate protective efficacy of multifarious NPs against promotion of Aβ aggregation, a narrow difference in size or surface chemistry may promote the aggregation process.[81–87] Therefore, to prevent such irreparable loss, a thorough understanding of NPs and their properties is warranted. Although 3D cell cultures appear to be appealing therapeutic arsenals in AD treatment, yet their molecular underpinnings involved in autonomous synapse generation remain elusive.[88] The primary impediments in transferrin-mediated BBB entry of NPs include synthesis procedure, stability, and immunological response.[38]

CONCLUSIONS AND FUTURE PERSPECTIVES

Despite the remarkable efficacy of currently available nanomaterials for the treatment of AD, their clinical translation has been hindered due to a few hitherto unresolved issues. However, voluminous studies conducted so far provide the understanding of structurally perturbed brain tissues due to the employment of nanomaterials.[9] Although BBB spanning target moieties or nanomaterials ameliorate AD symptoms significantly, yet the enhanced permeability of BBB may also have enhanced exposure to drugs and biomaterials, thus inducing pronounced toxicity to the brain.[9] Moreover, copious hitherto conducted neurotoxicity studies focused only on nanomaterials prepared by inorganic materials. Therefore, further neurotoxicity studies are warranted in drug delivering polymeric nanomaterials or lipid nanocarriers.[9] In addition, evaluation of brain toxicity is an extremely difficult task unlike the toxicity of other organs such as liver, heart, and kidney. Presence of extensively brain-targeted receptors such as LDL, insulin, and transferrin on the other organs is also a limitation in the development of appropriate NPs.[9] A thorough understanding of the underlying mechanisms of NPs functions and associated challenges *in vivo* such as protein corona etc. might be adequate to streamline appropriate nanomaterial-based therapeutic arsenals. Therefore, multitudinous therapeutic efficiency of multifarious biomaterials and the hitherto unresolved challenges leave an unending ambiguity for the neuroscientists whether the available materials are a boon or curse in the treatment of AD.

REFERENCES

1. Hung SY, Fu WM. Drug candidates in clinical trials for Alzheimer's disease. *J Biomed Sci*. 2017;24:47.
2. Maiti P, Dunbar GL. Comparative neuroprotective effects of dietary curcumin and solid lipid curcumin particles in cultured mouse neuroblastoma cells after exposure to Aβ42. *Int J Alzheimer's Dis*. 2017;2017:4164872.
3. Sarubbo F, Moranta D, Asensio VJ, Miralles A, Esteban S. Effects of resveratrol and other polyphenols on the most common brain age-related diseases. *Curr Med Chem*. 2017;24:4245–4266.
4. Monacelli F, Acquarone E, Giannotti C, Borghi R, Nencioni A. Vitamin C, aging and Alzheimer's disease. *Nutrients*. 2017;9.
5. Lakey-Beitia J, Doens D, Jagadeesh Kumar D, et al. Anti-amyloid aggregation activity of novel carotenoids: implications for Alzheimer's drug discovery. *Clin Interv Aging*. 2017;12:815–822.
6. Obulesu M, Venu R, Somashekhar R. Lipid peroxidation in Alzheimer's disease: emphasis on metal-mediated neurotoxicity. *Acta Neurol Scand*. 2011;124:295–301.
7. Obulesu M, Dowlathabad MR, Bramhachari PV. Carotenoids and Alzheimer's disease: an insight into therapeutic role of retinoids in animal models. *Neurochem Int*. 2011;59:535–541.
8. Frozza RL, Bernardi A, Hoppe JB, et al. Neuroprotective effects of resveratrol against Aβ administration in rats are improved by lipid-core nanocapsules. *Mol Neurobiol*. 2013;47:1066–1080.
9. Wong HL, Wu XY, Bendayan R. Nanotechnological advances for the delivery of CNS therapeutics. *Adv Drug Deliv Rev*. 2012;64:686–700.
10. Invernici G, Cristini S, Alessandri G, et al. Nanotechnology advances in brain tumors: the state of the art. *Recent Pat Anticancer Drug Discov*. 2011;6:58–69.
11. Wong HL, Bendayan R, Rauth AM, Li Y, Wu XY. Chemotherapy with anticancer drugs encapsulated in solid lipid nanoparticles. *Adv Drug Deliv Rev*. 2007;59:491–504.
12. Hajipour MJ, Santoso MR, Rezaee F, Aghaverdi H, Mahmoudi M, Perry G. Advances in Alzheimer's diagnosis and therapy: the implications of nanotechnology. *Trends Biotechnol*. 2017a;35:937–953.
13. Hajipour MJ, Ghasemi F, Aghaverdi H, et al. Sensing of Alzheimer's disease and multiple sclerosis using nano-bio interfaces. *J Alzheimers Dis*. 2017b;59:1187–1202.
14. Hadavi D, Poot AA. Biomaterials for the treatment of Alzheimer's disease. *Front Bioeng Biotechnol*. 2016;4:49.
15. Modi G, Pillay V, Choonara YE, Ndesendo VM, du Toit LC, Naidoo D. Nanotechnological applications for the treatment of neurodegenerative disorders. *Prog Neurobiol*. 2009;88:272–285.
16. Locatelli E, Franchini MC. Biodegradable PLGA-b-PEG polymeric nanoparticles: synthesis, properties, and nanomedical applications as drug delivery system. *J Nanopart Res*. 2012;14:1316.
17. Gao X, Qian J, Zheng S, et al. Overcoming the blood-brain barrier for delivering drugs into the brain by using adenosine receptor nanoagonist. *ACS Nano*. 2014;8:3678–3689.
18. Saraiva C, Praca C, Ferreira R, Santos T, Ferreira L, Bernardino L. Nanoparticle-mediated brain drug delivery: overcoming blood-brain barrier to treat neurodegenerative diseases. *J Control Release*. 2016;235:34–47.
19. Choi CHJ, Alabi CA, Webster P, Davis ME. Mechanism of active targeting in solid tumors with transferrin-containing gold nanoparticles. *Proc Natl Acad Sci USA*. 2010;107:1235–1240.
20. Wiley DT, Webster P, Gale A, Davis ME. Transcytosis and brain uptake of transferrin-containing nanoparticles by tuning avidity to transferrin receptor. *Proc Natl Acad Sci USA*. 2013;110:8662–8667.
21. Kong SD, Lee J, Ramachandran S, et al. Magnetic targeting of nanoparticles across the intact blood-brain barrier. *J Contr Release*. 2012;164:49–57.
22. Fleegal-DeMotta MA, Doghu S, Banks WA. Angiotensin II modulates BBB permeability via activation of the AT(1) receptor in brain endothelial cells. *J Cerebr Blood Flow Metabol*. 2009;29:640–647.
23. Qin LJ, Gu YT, Zhang H, Xue YX. Bradykinin-induced blood–tumor barrier opening is mediated by tumor necrosis factor-alpha. *Neurosci Lett*. 2009;450:172–175.
24. Sood RR, Taheri S, Candelario-Jalil E, Estrada EY, Rosenberg GA. Early beneficial effect of matrix metalloproteinase inhibition on blood–brain barrier permeability as measured by magnetic resonance imaging countered by impaired long-term recovery after stroke in rat brain. *J Cerebr Blood Flow Metabol*. 2008;28:431–438.

25. van der Sandt IC, Gaillard PJ, Voorwinden HH, de Boer AG, Breimer DD. P-glycoprotein inhibition leads to enhanced disruptive effects by antimicrotubule cytostatics at the *in vitro* blood–brain barrier. *Pharm Res.* 2001;18:587–592.

26. Giordano C, Albani D, Gloria A, et al. Nanocomposites for neurodegenerative diseases: hydrogel-nanoparticle combinations for a challenging drug delivery. *Int J Artif Organs.* 2011;34:1115–1127.

27. Sun D, Li N, Zhang W, et al. Design of PLGA-functionalized quercetin nanoparticles for potential use in Alzheimer's disease. *Colloids Surf B Biointerfaces.* 2016;148:116–129.

28. Amin FU, Shah SA, Badshah H, Khan M, Kim MO. Anthocyanins encapsulated by PLGA@PEG nanoparticles potentially improved its free radical scavenging capabilities via p38/JNK pathway against Aβ1-42-induced oxidative stress. *J Nanobiotechnol.* 2017;15:12.

29. Smith A, Giunta B, Bickford PC, Fountain M, Tan J, Shytle RD. Nanolipidic particles improve the bioavailability and alpha-secretase inducing ability of epigallocatechin-3-gallate (EGCG) for the treatment of Alzheimer's disease. *Int J Pharm.* 2009;389:207–212.

30. Alyautdin R, Gothier D, Petrov V, Kharkevich D, Kreuter J. Analgesic activity of the hexapeptide dalargin adsorbed on the surface of polysorbate 80-coated poly(butyl cyanoacrylate) nanoparticles. *Eur J Pharm Biopharm.* 1995;41:44–48.

31. Gao K, Jiang X. Influence of particle size on transport of methotrexate across blood–brain barrier by polysorbate 80-coated poly(butylcyanoacrylate) nanoparticles. *Int J Pharm.* 2006;310:213–219.

32. Kim HR, Andrieux K, Gil S, et al. Translocation of poly(ethylene glycol-co-hexadecyl) cyanoacrylate nanoparticles into rat brain endothelial cells: role of apolipoproteins in receptor-mediated endocytosis. *Biomacromolecules.* 2007;8:793–799.

33. Tian XH, Lin XN, Wei F, et al. Enhanced brain targeting of temozolomide in polysorbate-80 coated poly(butylcyanoacrylate) nanoparticles. *Int J Nanomed.* 2011;6:445–452.

34. Alyaudtin RN, Reichel A, Lobenberg R, Ramge P, Kreuter J, Begley DJ. Interaction of poly(butylcyanoacrylate) nanoparticles with the blood–brain barrier *in vivo* and *in vitro.* *J Drug Target.* 2001;9:209–221.

35. Rempe R, Cramer S, Huwel S, Galla HJ. Transport of poly(n-butylcyano-acrylate) nanoparticles across the blood–brain barrier *in vitro* and their influence on barrier integrity. *Biochem Biophys Res Commun.* 2011;406:64–69.

36. Das D, Lin S. Double-coated poly(butylcynanoacrylate) nanoparticulate delivery systems for brain targeting of dalargin via oral administration. *J Pharm Sci.* 2005;94:1343–1353.

37. Wilson B. Brain targeting PBCA nanoparticles and the blood-brain barrier. *Nanomedicine.* 2009;4:499–502.

38. Yin T, Yang L, Liu Y, Zhou X, Sun J, Liu J. Sialic acid (SA)-modified selenium nanoparticles coated with a high blood-brain barrier permeability peptide-B6 peptide for potential use in Alzheimer's disease. *Acta Biomater.* 2015;25:172–183.

39. Kulkarni PV, Roney CA, Antich PP, Bonte FJ, Raghu AV, Aminabhavi TM. Quinoline-n-butylcyanoacrylate-based nanoparticles for brain targeting for the diagnosis of Alzheimer's disease. *Wiley Interdiscip Rev Nanomed Nanobiotechnol.* 2010;2:35–47.

40. Hao C, Wang W, Wang S, Zhang L, Guo Y. An overview of the protective effects of chitosan and Acetylated chitosan oligosaccharides against neuronal disorders. *Mar Drugs.* 2017;15:E89.

41. Wen MM, El-Salamouni NS, El-Refaie WM, et al. Nanotechnology-based drug delivery systems for Alzheimer's disease management: technical, industrial, and clinical challenges. *J Control Release.* 2017;245:95–107.

42. Salles GN, Pereira FA, Pacheco-Soares C, et al. A novel bioresorbable device as a controlled release system for protecting cells from oxidative stress from Alzheimer's disease. *Mol Neurobiol.* 2017;54.

43. Marei HE, Elnegiry AA, Zaghloul A, et al. Nanotubes impregnated human olfactory bulb neural stem cells promote neuronal differentiation in trimethyltin-induced neurodegeneration rat model. *J Cell Physiol.* 2017;232:3586–3597.

44. Sacramento AS, Moreira FTC, Guerreiro JL, Tavares AP, Sales MGF. Novel biomimetic composite material for potentiometric screening of acetylcholine, a neurotransmitter in Alzheimer's disease. *Mater Sci Eng C Mater Biol Appl.* 2017;79:541–549.

45. Zhang L, Liu F, Sun X, et al. Engineering carbon nanotube fiber for real-time quantification of ascorbic acid levels in a live rat model of Alzheimer's disease. *Anal Chem.* 2017;89:1831–1837.

46. Zhu G, Lee HJ. Electrochemical sandwich-type biosensors for α-1 antitrypsin with carbon nanotubes and alkaline phosphatase labeled antibody-silver nanoparticles. *Biosens Bioelectron.* 2017;89:959–963.

47. Okada M, Nakai A, Hara ES, Taguchi T, Nakano T, Matsumoto T. Biocompatible nanostructured solid adhesives for biological soft tissues. *Acta Biomater.* 2017;57:404–413.

48. Giordano C, Albani D, Gloria A, et al. Multidisciplinary perspectives for Alzheimer's and Parkinson's diseases: hydrogels for protein delivery and cell-based drug delivery as therapeutic strategies. *Int J Artif Organs.* 2009;32:836–850.

49. Murphy AR, Laslett A, O'Brien CM, Cameron NR. Scaffolds for 3D *in vitro* culture of neural lineage cells. *Acta Biomater.* 2017;54:1–20.

50. Jacob RS, Ghosh D, Singh PK, et al. Self healing hydrogels composed of amyloid nano fibrils for cell culture and stem cell differentiation. *Biomaterials.* 2015;54:97–105.

51. Boridy S, Takahashi H, Akiyoshi K, Maysinger D. The binding of pullulan modified cholesteryl nanogels to Abeta oligomers and their suppression of cytotoxicity. *Biomaterials.* 2009;30:5583–5591.

52. Katz JS, Burdick JA. Hydrogel mediated delivery of trophic factors for neural repair. *Wiley Interdiscip Rev Nanomed Nanobiotechnol.* 2009;1:128–139.
53. Nahar M, Dutta T, Murugesan S, et al. Functional polymeric nanoparticles: an efficient and promising tool for active delivery of bioactives. *Crit Rev Ther Drug Carrier Syst.* 2006;23:259–318.
54. Sundaram RK, Kasinathan C, Stein S, Sundaram P. Detoxification depot for beta-amyloid peptides. *Curr Alzheimer Res.* 2008;5:26–32.
55. Zhong Y, Bellamkonda RV. Biomaterials for the central nervous system. *J R Soc Interface.* 2008;5:957–975.
56. Gloria A, De Santis R, Ambrosio L. Polymer-based composite scaffolds for tissue engineering. *J Appl Biomater Biomech.* 2010;8:57–67.
57. Gloria A, Russo T, De Santis R, Ambrosio L. 3D fiber deposition technique to make multifunctional and tailor-made scaffolds for tissue engineering applications. *J Appl Biomater Biomech.* 2009;7:141–152.
58. Sionkowska A. Current research on the blends of natural and synthetic polymers as new biomaterials: review. *Prog Polym Sci.* 2011;36:1254–1276.
59. Choi SH, Tanzi RE. iPSCs to the rescue in Alzheimer's research. *Cell Stem Cell.* 2012;10:235–236.
60. Kim YH, Choi SH, D'Avanzo C, et al. A 3D human neural cell culture system for modeling Alzheimer's disease. *Nat Protoc.* 2015;10:985–1006.
61. Saha K, Jaenisch R. Technical challenges in using human induced pluripotent stem cells to model disease. *Cell Stem Cell.* 2009;5:584–595.
62. Young JE, Goldstein LSB. Alzheimer's disease in a dish: promises and challenges of human stem cell models. *Hum Mol Genet.* 2012;21:R82–R89.
63. Choi SH, Kim YH, Hebisch M, et al. A three-dimensional human neural cell culture model of Alzheimer's disease. *Nature.* 2014;515:274–278.
64. Lauzon MA, Daviau A, Marcos B, Faucheux N. Nanoparticle-mediated growth factor delivery systems: a new way to treat Alzheimer's disease. *J Control Release.* 2015;206:187–205.
65. Di Stefano A, Iannitelli A, Laserra S, Sozio P. Drug delivery strategies for Alzheimer's disease treatment. *Expert Opin Drug Deliv.* 2011;8:581–603.
66. Herran E, Perez-Gonzalez R, Igartua M, Pedraz JL, Carro E, Hernandez RM. Enhanced hippocampal neurogenesis in APP/Ps1 mouse model of Alzheimer's disease after implantation of VEGF-loaded PLGA nanospheres. *Curr Alzheimer Res.* 2015;12:932–940.
67. Faustino C, Rijo P, Reis CP. Nanotechnological strategies for nerve growth factor delivery: therapeutic implications in Alzheimer's disease. *Pharmacol Res.* 2017;120:68–87.
68. Thorne RG, Frey II WH. Delivery of neurotrophic factors to the central nervous system: pharmacokinetic considerations. *Clin Pharmacokinet.* 2001;40:907–946.
69. Thoenen H, Sendtner M. Neurotrophins: from enthusiastic expectations through sobering experiences to rational therapeutic approaches. *Nat Neurosci.* 2002;5:1046–1050.
70. Lee JS, Go DH, Bae JW, Lee SJ, Park KD. Heparin conjugated polymeric micelle for long-term delivery of basic fibroblast growth factor. *J Contr Release.* 2007;117:204–209.
71. Chen Y, Pan C, Xuan A, et al. Treatment efficacy of NGF nanoparticles combining neural stem cell transplantation on Alzheimer's disease model rats. *Med Sci Monit.* 2015;21:3608–3615.
72. Obulesu M, Jhansilakshmi M. Neuroprotective role of nanoparticles against Alzheimer's disease. *Curr Drug Metab.* 2016;17:142–149.
73. Cherney RA, Atwood CS, Xilinas ME, et al. Treatment with a copper-zinc chelator markedly and rigidly inhibits beta-amyloid accumulation in Alzheimer's disease transgenic mice. *Neuron.* 2001;30:665–676.
74. Cui Z, Lockman P, Atwood R, et al. Novel d-penicillamine carrying nanoparticles for metal chelation therapy in Alzheimer's and other CNS diseases. *Eur J Pharm Biopharm.* 2005;59:263–272.
75. Liu G, Men P, Perry G, Smith MA. Development of iron chelator nanoparticle conjugates as potential therapeutic agents for Alzheimer disease. *Prog Brain Res.* 2010;180:97–108.
76. Cui Z, Bu W, Fan W, et al. Sensitive imaging and effective capture of Cu(2+): towards highly efficient theranostics of Alzheimer's disease. *Biomaterials.* 2016;104:158–167.
77. Liu H, Dong X, Liu F, Zheng J, Sun Y. Iminodiacetic acid-conjugated nanoparticles as a bifunctional modulator against Zn2+-mediated amyloid β-protein aggregation and cytotoxicity. *J Colloid Interface Sci.* 2017;505:973–982.
78. Mufamadi MS, Choonara YE, Kumar P, et al. Surface-engineered nanoliposomes by chelating ligands for modulating the neurotoxicity associated with β-amyloid aggregates of Alzheimer's disease. *Pharm Res.* 2012;29:3075–3089.
79. Billi F, Campbell P. Nanotoxicology of metal wear particles in total joint arthroplasty: a review of current concepts. *J Appl Biomater Biomech.* 2010;8:1–6.
80. Kunzmann A, Andersson B, Thurnherr T, Krug H, Scheynius A, Fadeel B. Toxicology of engineered nanomaterials: focus on biocompatibility, biodistribution and biodegradation. *Biochim Biophys Acta.* 2011;1810:361–373.
81. Cabaleiro-Lago C, Quinlan-Pluck F, Lynch I, Dawson KA, Linse S. Dual effect of amino modified polystyrene nanoparticles on amyloid β protein fibrillation. *ACS Chem Neurosci.* 2010;1:279–287.
82. Ghavami M, Rezaei M, Ejtehadi R, et al. Physiological temperature has a crucial role in amyloid beta in the absence and presence of hydrophobic and hydrophilic nanoparticles. *ACS Chem Neurosci.* 2013;4:375–378.
83. Ma Q, Wei G, Yang X. Influence of Au nanoparticles on the aggregation of amyloid-β-(25–35) peptides. *Nanoscale.* 2013;5:10397–10403.
84. Mahmoudi M, Quinlan-Pluck F, Monopoli MP, et al. Influence of the physiochemical properties of superparamagnetic iron oxide nanoparticles on amyloid β protein fibrillation in solution. *ACS Chem Neurosci.* 2013;4:475–485.

85. Moore KA, Pate KM, Soto-Ortega DD, et al. Influence of gold nanoparticle surface chemistry and diameter upon Alzheimer's disease amyloid-β protein aggregation. *J Biol Eng.* 2017;11:5.

86. Saraiva AM, Cardoso I, Pereira MC, et al. Controlling amyloid-β peptide(1–42) oligomerization and toxicity by fluorinated nanoparticles. *Chembiochem.* 2010;11:1905–1913.

87. Wu WH, Sun X, Yu YP, et al. TiO$_2$ nanoparticles promote β-amyloid fibrillation *in vitro*. *Biochem Biophys Res Commun.* 2008;373:315–318.

88. Puschmann TB, de Pablo Y, Zanden C, Liu J, Pekny M. A novel method for three-dimensional culture of central nervous system neurons. *Tissue Eng Part C Methods.* 2014;20:485–492.

FURTHER READING

1. Pardridge WM. Drug targeting to the brain. *Pharm Res.* 2007;24:1733–1744.

2. Tsaioun K, Bottlaender M, Mabondzo A, Alzheimer's Drug Discovery Foundation. ADDME-Avoiding Drug Development Mistakes Early: central nervous system drug discovery perspective. *BMC Neurol.* 2009;9:S1.

CHAPTER 10

Alzheimer Therapeutics: Pros and Cons

ABSTRACT

Since the discovery of Alzheimer's disease (AD) in 1906 its pathology has been extensively studied and multifarious therapeutics were used to overcome disease symptoms. A few therapeutics hold good promise for AD although manifold therapeutics failed at clinical trials. This chapter throws light on the pros and cons of AD therapeutics. Multifarious and novel therapeutics discussed in this book have been summarized in this chapter.

KEYWORDS

Alzheimer's disease; Antioxidants; Drugs; Nanoparticles.

INTRODUCTION

Complex etiology of Alzheimer's disease (AD), biological riddles entailed in diagnosis and blood-brain barrier (BBB) limit the therapeutic effect of numerous compounds significantly (Chapters 1 and 5). With a view to overcoming challenges associated with low-molecular-weight compounds like drugs and bioactive compounds, a plethora of nanoparticles were designed and extensively used in AD treatment. The following are the merits and demerits of hitherto unraveled methods. As AD is a progressive neurodegenerative disorder, a multidimensional diagnostic method is a prerequisite.[1] Although a few innovative protein biomarkers such as APLP1 and SPP1 presented satisfactory specificity in AD diagnosis, a few studies produced opposing results, thus leaving a potential uncertainty.[2] Although there is apposite use of MRS in AD diagnosis, a few shortcomings such as prolonged time consumption for data attaining and the presence of analogous hardware in the clinic hinder its success.[1] Amazing collection of data from multidimensional diagnostic tools and contribution of artificial intelligence by a few neuroscientists are estimated to lay an important milestone to reorganize AD diagnosis perfectly and intensely.[1,3]

Neuroinflammation and Cytokines

Neuroinflammation plays a key role in AD pathology.[4] Hence, numerous therapies targeted at neuroinflammation show extensive ability presently and appeal the attention of neuroscientists.[5–8] However, there is also huge scarcity in neuroinflammation-targeted therapies. To cater to the needs, a tough BBB traversing compound N-((5-(3-(1-benzylpiperidin-4-yl)propoxy)-1-methyl-1H-indol-2-yl)methyl)-N-methylprop-2-yn-1-amine (ASS234) was studied satisfactorily.[9–16] Nevertheless, no satisfactory studies were instigated on the toxicity profile of this compound. Therefore, the influence on genes was investigated to fill the gap. ASS234 positively blocked the inflammation in lipopolysaccharide (LPS)-activated RAW 264.7 macrophages. To recognize the fundamental mechanism of neuroprotection gene expression of interleukin-6 (IL6), IL1β, TNF-α, TNFR1, NF-κB, IL-10, and TGF-β was evaluated in SH-SY5Y cells.[8,17–20] Since they play a crucial role in stimulation of inflammatory response, their regulation is a correct avenue to overcome inflammation.[8] NF-κB was found to regulate IL-6, TNF-α, and IL-1β, which are accountable for BBB injury.[8] ASS234 being a strong inhibitor of this mechanistic pathway reduced BBB damage in AD patients.[8] ASS234 treatment also upregulated IL-10 and TGF-β, which in turn instigated an antiapoptotic pathway and accrued cell survival.[8]

Demerits

Despite noteworthy influence of gene therapy in progress of AD therapeutics, a few shortcomings are to be considered. Regrettably, multiple immunotherapeutic studies directed against Aβ returned ineffective results.[21–23] Despite outstanding therapeutic efficacy of a few strategies beset against neuroinflammation in AD, dearth in molecular understanding has been a problem in building them as strong therapeutic resources. In order to combat these issues, a multitargeted small molecule, ASS234, synthesized and widely used to downregulate inflammatory genes, shows satisfactory potential in AD therapeutics presently.[8] Consequently,

Alzheimer's Disease Theranostics. https://doi.org/10.1016/B978-0-12-816412-9.00010-0
Copyright © 2019 Elsevier Inc. All rights reserved.

there is an escalating need to enhance emphasis on this compound to project it as a clinically appropriate tool to overcome AD.

Antioxidants

Growing evidence suggests that melatonin (N-acetyl-5-methoxytryptamine) which has a role in circadian rhythm also alleviates ROS and RNS generation and aggravates the antioxidant enzyme activity.[24,25] Moreover, melatonin can decline Aβ load, tau hyperphosphorylation, and kainic acid–induced microglial and astroglial reactions.[25–28] Having these noticeable therapeutic benefits, melatonin has widely been used as an amenable scaffold to combine diverse compounds such as melatonin donepezil hybrids.[25,29] Multiple lines of evidence proposes a potential pathological link between T2D pathology and AD, and T2D doubles the AD risk.[30,31] Voluminous data disclose that tough antioxidant, antiglucosidase, and anticholinesterase activities of methanolic extracts and chloroform fractions of *Buchanania axillaris, Hemidesmus indicus,* and *Rus mysorensis* expressively improve AD and type 2 diabetes (T2D) pathology.[31]

Shortcomings

Despite compatibility and noteworthy therapeutic potential of antioxidants, they exhibited inadequate efficacy in AD treatment. A few explanations given for low efficacy include scarce dose and insufficient therapy interval.[25,32–34] With a tough antioxidant and anti-inflammatory efficacy, ferulic acid has widely been used as a scaffold to target AD by tacrine-ferulic acid hybrids.[25,35–38]

DRUG MECHANISM/GENE MODIFICATION

Currently, mounting evidence has shown that ceftriaxone, an antibiotic with sufficient neuroprotective efficacy, revealed an insightful influence on genes. Accordingly, ceftriaxone offered amazing decrease in mRNA levels of *Bace1* (encode) and *Ace2* in the hypothalamus and Aktb in the frontal cortex in OXYS rats.[39] Also, higher Mme, Ide, and Epo mRNA levels in amygdala and the levels of Ece1 and Aktb in the striatum were also witnessed.[39] Since Aβ formation is contingent upon APP cleavage by BACE 1 and its scavenging by mechanisms such as proteolytic breakdown, transport and aggregation, the comparable gene modulation strategies were used.[39–43]

Neprilysin

Neprilysin (NEP) strongly degenerates both monomeric and oligomeric forms.[44] Consequently, NEP decline has also been involved in AD pathology.[44] Therefore, NEP distribution and expression improvement has been chosen as a potential AD therapeutic currently.[44,45] NEP decline revealed enriched levels of Aβ40 and Aβ42 in APP transgenic mice, which were abnormally reduced by the delivery of NEP expressing viral vectors.[44,46,47] Excitingly, PGRN delivered through AAV vectors showed increased expression, thus improving NEP expression followed by Aβ plaque deterioration.[44]

Limitations

Despite extraordinary growth in viral vector therapeutics against AD, their efficacy has to be validated.[48] Although viral vector therapeutics amended symptoms of numerous diseases, yet a few shortcomings such as induction of autoimmunity and initiation of pre-existing immunity against viral vector impede their success.[48] While AAV9 has been known to utilize galactose binding and absorptive endocytosis to ferry BBB, yet further studies are necessary to substantiate these mechanisms.[49]

Nanotechnology

Nanoparticles (NPs) crop up as trustworthy theranostic tools in AD treatment due to a few substantial characteristics such as the possibility to surface functionalize with appropriate ligands, notable efficacy to span biological membranes, targeted delivery of therapeutic compounds, and improved permeation and retention ability.[50,51] Of the diverse polymers, poly(ethylene) glycol (PEG) has been found to be biocompatible and improve systemic circulation and is widely used in the synthesis of NPs.[51] Various Nps targeted against AD comprise Aβ-targeted Nps and Tau-targeted NPs (see Fig. 1.1).

Wealth of studies also indicated that magnetic core-plasmonic coat nanomaterial linked to hybrid graphene oxide with tau and Aβ assist in specific screening of tau and Aβ in patients.[51] Another study presented that light-emitting quantum dots reach the target site in the brain and ameliorate Aβ-prompted neuroinflammation and oxidative stress by laser therapy.[51,52] Metal dyshomeostasis also plays a central role in AD pathology by manipulating the key enzymes responsible for Aβ generation and scavenging.[51] To circumvent the metal intoxication, metal chelation therapy has been widely used.[53] Recently developed ultrathin graphitic phase carbon nitride (g-C_3N_4) removed copper (Cu^{2+}) ions and reduced Aβ fibrillation.[54] H_2O_2-sensitive silica nanocarriers were also synthesized to undertake targeted delivery of metal chelators to the H_2O_2 generating cells.[51,55]

Gold Nanoparticles

Betzer et al. (2017) developed insulin-linked gold NPs (INS-GNPs) and achieved targeting to the particular brain regions. According to their study, INS-GNPs with 20 nm productively crossed BBB via insulin receptors on BBB and acted as computed tomography contrast agents, thus aiding noninvasive identification of particle accumulation (*in vivo*).[56,57] GNPs used in this study offer incredible features such as size compliance, enhanced circulation time, stability, and biosafety.[56,57,59] Accordingly, it can be agreed that the GNP can form a vigorous core which can enable diverse ligand conjugation and achieve many therapeutic effects.[56]

Liposomes

Liposomes are pliable DDS in brain targeting since their phospholipid bilayer mimics the physiological cell membrane.[60] To achieve distinct brain targeting, surface functionalization of liposomes with ligands like lactoferrin, transferrin, glutathione, and glucose was carried out and targeted to treat AD.[60] Rip et al. (2014) developed glutathione functionalized PEGylated liposomes and analyzed their brain distribution using carboxyfluorescein, a fluorescent particle (*in vivo*).[61]

Magnetic Nanoparticles

Magnetic nanocontainers with electromagnetic fields of 28 mT (0.43 T/m) and 79.8 mT (1.39 T/m) were designed and administered via tail vein which showed active BBB passage, thus improving AD symptoms in mice.[62] In another study, PEG and mono-dispersed nitrodopamine-associated magnetic NPs were prepared, which can modify their surface via carboxylation and can further bind to AβO-specific antibodies.[63] Interestingly, osmotin-loaded magnetic NPs were designed, which unveiled superior BBB spanning and improvement of synaptic defects and Aβ aggregation.[64] In this study, fluorescent carboxyl magnetic Nile Red particles (FMNPs) were synthesized and driven even to the hippocampus and cortex regions in mice through functionalized magnetic field (FMF).[64]

Cerium Oxide Nanoparticles

TPP-linked ceria (CeO_2) NPs act as vigorous ROS scavengers in mitochondria by transitioning between Ce3+ and Ce4+ oxidation states and ultimately diminish cell death in 5XFAD AD model mice.[65] These NPs inhibit reactive gliosis and mitochondrial dysfunction in mice.[65] These NPs presented stable hydrodynamic diameter and colloidal stability when incubated with phosphate-buffered saline (PBS) and Dulbeccos

modified Eagle's Medium (DMEM) and 10% fetal bovine serum (FBS).[65]

In another study, Ce3+ and Ce4+ transition property has been used to treat mitochondrial fission. Hence, nanoceria were used which were deeply localized in mitochondrial outer membrane and opposed Aβ and peroxynitrite instigated mitochondrial fission.[66] Peroxynitrite generally plays an essential role in Aβ aggregation and neurofibrillary tangle (NFT) formation.[66–69] This effect of mitochondrial fission and cell death has been prompted through activation of dynamin-related-protein1 (DRP1), a gigantic GTPase that facilitates mitochondrial fission by hitherto unidentified molecular mechanisms.[66] Nevertheless, DRP1 serine 616 phosphorylation plays a possible role. Cerium oxide NPs successfully eradicate superoxide anions, hydrogen peroxide, and peroxynitrite.[66] They also intensely inhibit ROS and reactive nitrogen species (RNS).[66]

LIMITATIONS OF MITOCHONDRIA-TARGETED THERAPEUTICS

Despite scarcity of studies on molecular basis of mitochondrial dysfunction, surplus of former studies uncovered that interaction of Aβ with a few mitochondrial proteins like cyclophilin D, alcohol dehydrogenase, and ATP synthase promote expressively to ROS generation.[65,70–72]

Although SWCNTs embrace a substantial promise for AD therapeutics, yet abundant studies showed their cytotoxicity.[73–86]

DEMERITS OF BIOMATERIALS

NPs which play a pivotal role in circumventing not only AD but also numerous other diseases induce considerable systemic toxicity.[87–89] Additionally, the tremendously expensive therapeutic strategies used to treat brain diseases accounts to US$ 100 million–US$ 1 billion prior to reaching the market. Nevertheless, only 3%–5% of the therapeutics reaches the market due to the BBB obstacle.[90] In spite of the satisfactory protective ability of different NPs against advancement of Aβ aggregation, a slight difference in size or surface chemistry may support the aggregation process.[91–97] Consequently, to avert such long-lasting loss, a detailed understanding of NPs and their properties is acceptable. Although 3D cell cultures seem to be interesting therapeutic resources in AD treatment, yet their molecular mechanisms entailed in autonomous synapse generation remains obscure.[98] The chief obstructions in transferrin-facilitated BBB entry of NPs comprise synthesis procedure, stability, and immunological response.[99]

CONCLUSIONS AND FUTURE PERSPECTIVES

BBB plays a critical role in regular physiology of brain and offers amazing protection by avoiding the entry of pathogens. Conversely, it has also been regarded a key barrier and a biological puzzle in the treatment of neurodegenerative disorders like AD. Panoply of studies targeting conventional treatment options like antioxidant, anti-inflammatory, and cholinesterase inhibition therapies against AD returned inadequate success. In order to overcome these issues, multiple nanotechnological methods have been widely used. Despite principal studies and their ability in improvement of drug delivery through BBB, their clinical translation has been a difficult task till date. Besides, a few strategies used to achieve BBB passing may show forthcoming antagonistic effects such as permeation of toxic components into the brain, etc. Hence, significant future nanotechnological studies are warranted to achieve the effective BBB and AD therapy.

Even though currently available nanomaterials show significant therapeutic efficacy in the treatment of AD, their clinical translation has been delayed due to a few yet unanswered issues. Nevertheless, ample studies directed so far offer the understanding of structurally agitated brain tissues due to the use of nanomaterials.[100] Although BBB spanning target moieties or nanomaterials improve AD symptoms considerably, yet the improved perviousness of BBB may also have improved exposure to drugs and biomaterials, thus inducing noticeable toxicity to the brain.[100] Besides, plentiful hitherto-directed neurotoxicity studies focused only on nanomaterials prepared by inorganic materials. Consequently, additional neurotoxicity studies are reasonable in drug delivering polymeric nanomaterials or lipid nanocarriers.[100] Moreover, assessment of brain toxicity is an exceptionally hard task unlike the toxicity of other organs like liver, heart, and kidney. Occurrence of widely brain-targeted receptors such as LDL, insulin, and transferrin on the other organs is also a constraint in the design of suitable NPs.[100] A detailed insight into the underlying mechanisms of NP functions, related to challenges *in vivo* such as protein corona, etc. might be acceptable to restructure apposite nanomaterial-based therapeutic tools. Therefore, infinite therapeutic efficiency of manifold biomaterials and the hitherto unsettled challenges rest an unending uncertainty for the neuroscientists, whether the existing materials are a blessing or curse in the treatment of AD.

REFERENCES

1. Mandal PK, Shukla D. Brain metabolic, structural, and behavioral pattern learning for early predictive diagnosis of Alzheimer's disease. *J Alzheimers Dis.* 2018;63:935–939.
2. Begcevic I, Brinc D, Brown M, Martinez-Morillo E, Goldhardt O, Grimmer T. Brain-related proteins as potential CSF biomarkers of Alzheimer's disease:targeted mass spectrometry approach. *J Proteomics.* 2018;182:12–20.
3. Shen D, Wu G, Suk HI. Deep learning in medical image analysis. *Annu Rev Biomed Eng.* 2017;19:221–248.
4. Obulesu M, Jhansilakshmi M. Neuroinflammation in Alzheimer's disease: an understanding of physiology and pathology. *Int J Neurosci.* 2014;124:227–235.
5. Lyman M, Lloyd DG, Ji X, Vizcaychipi MP, Ma D. Neuroinflammation: the role and consequences. *Neurosci Res.* 2014;79:1–12.
6. Ardura-Fabregat A, Boddeke E, Boza-Serrano A, et al. Targeting neuroinflammation to treat Alzheimer's disease. *CNS Drugs.* 2017;31:1057–1082.
7. Dansokho C, Heneka MT. Neuroinflammatory responses in Alzheimer's disease. *J Neural Transm.* 2018;125:771.
8. del Pino J, Marco-Contelles J, Lopez-Munoz F, Romero A, Ramos E. Neuroinflammation signaling modulated by ASS234 a multitarget small molecule for Alzheimer's disease therapy. *ACS Chem Neurosci.* 2018. [in press].
9. Bolea I, Gella A, Monjas L, et al. Multipotent, permeable drug ASS234 inhibits Abeta aggregation, possesses antioxidant properties and protects from Abeta-induced apoptosis *in vitro. Curr Alzheimer Res.* 2013;10:797–808.
10. Esteban G, Van Schoors J, Sun P, et al. *In-vitro* and *in-vivo* evaluation of the modulatory effects of the multitarget compound ASS234 on the monoaminergic system. *J Pharm Pharmacol.* 2017;69:314–324.
11. Serrano MP, Herrero-Labrador R, Futch HS, et al. The proof-of-concept of ASS234: peripherally administered ASS234 enters the central nervous system and reduces pathology in a male mouse model of Alzheimer disease. *J Psychiatry Neurosci.* 2017;42:59–69.
12. Marco-Contelles J, Unzeta M, Bolea I, Esteban G, Ramsay RR, Romero A. ASS234, as a new multi-target directed propargylamine for Alzheimer's disease therapy. *Front Neurosci.* 2016;10:294.
13. Ramos E, Romero A, Marco-Contelles J, Del Pino J. Up-regulation of antioxidant enzymes by ASS234, a multitarget directed propargylamine for Alzheimer's disease therapy. *CNS Neurosci Ther.* 2016;22:799–802.
14. Del Pino J, Ramos E, Aguilera OM, Marco-Contelles J, Romero A. Wnt signaling pathway, a potential target for Alzheimer's disease treatment, is activated by a novel multitarget compound ASS234. *CNS Neurosci Ther.* 2014;20:568–570.
15. Esteban G, Allan J, Samadi A, et al. Kinetic and structural analysis of the irreversible inhibition of human monoamine oxidases by ASS234, a multi-target compound designed for use in Alzheimer's disease. *Biochim Biophys Acta.* 2014;1844:1104–1110.

16. Bolea I, Juarez-Jimenez J, de Los Rios C, et al. Synthesis, biological evaluation, and molecular modeling of donepezil and N-[(5-(benzyloxy)-1-methyl-1H-indol-2-yl)methyl]-N-methylprop-2-yn-1-amine hybrids as new multipotent cholinesterase/monoamine oxidase inhibitors for the treatment of Alzheimer's disease. *J Med Chem.* 2011;54:8251–8270.
17. Siren AL, McCarron R, Wang L, et al. Proinflammatory cytokine expression contributes to brain injury provoked by chronic monocyte activation. *Mol Med.* 2001;7:219–229.
18. Medeiros R, Prediger RD, Passos GF, et al. Connecting TNF-alpha signaling pathways to iNOS expression in a mouse model of Alzheimer's disease: relevance for the behavioral and synaptic deficits induced by amyloid beta protein. *J Neurosci.* 2007;27:5394–5404.
19. Su F, Bai F, Zhang Z. Inflammatory cytokines and Alzheimer's disease: a review from the perspective of genetic polymorphisms. *Neurosci Bull.* 2016;32:469–480.
20. Zheng Y, Fang W, Fan S, et al. Neurotropin inhibits neuroinflammation via suppressing NF-kappaB and MAPKs signaling pathways in lipopolysaccharide-stimulated BV2 cells. *J Pharmacol Sci.* 2018;136:242.
21. Salloway S, Sperling R, Fox NC, et al. Bapineuzumab 301 and 302 Clinical Trial Investigators. Two phase 3 trials of bapineuzumab in mild-to-moderate Alzheimer's disease. *N Engl J Med.* 2014;370:322–333.
22. Doody RS, Thomas RG, Farlow M, et al. Alzheimer's disease cooperative study steering committee; solanezumab study group. Phase 3 trials of solanezumab for mild-to-moderate Alzheimer's disease. *N Engl J Med.* 2014;370:311–321.
23. Gyorgy B, Loov C, Zaborowski MP, et al. CRISPR/Cas9 mediated disruption of the Swedish APP allele as a therapeutic approach for early-onset Alzheimer's disease. *Mol Ther Nucleic Acids.* 2018;11:429–440.
24. Miller E, Morel A, Saso L, Saluk J. Melatonin redox activity. Its potential clinical applications in neurodegenerative disorders. *Curr Top Med Chem.* 2015;15:163–169.
25. Mezeiova E, Spilovska K, Nepovimova E, et al. Profiling donepezil template into multipotent hybrids with antioxidant properties. *J Enzym Inhib Med Chem.* 2018;33:583–606.
26. Lahiri DK. Melatonin affects the metabolism of the beta amyloid precursor protein in different cell types. *J Pineal Res.* 1999;26:137–146.
27. Rosales-Corral S, Tan DX, Reiter RJ, et al. Orally administered melatonin reduces oxidative stress and proinflammatory cytokines induced by amyloid-beta peptide in rat brain: a comparative, *in vivo* study versus vitamin C and E. *J Pineal Res.* 2003;35:80–84.
28. Li XC, Wang ZF, Zhang JX, Wang Q, Wang JZ. Effect of melatonin on calyculin A-induced tau hyperphosphorylation. *Eur J Pharmacol.* 2005;510:25–30.
29. Ramos E, Romero A, Marco-Contelles J, Del Pino J. Up-regulation of antioxidant enzymes by ASS234, a multi-target directed propargylamine for alzheimer's disease therapy. *CNS Neurosci Ther.* 2016;22:799–802.
30. Giuseppe V, Stephanie JF, Ralph NM. The role of type 2 diabetes in neurodegeneration. *Neurobiol Dis.* 2015;84:22–38.
31. Penumala M, Zinka RB, Shaik JB, Mallepalli SKR, Ramakrishna V, et al. Phytochemical profiling and *in vitro* screening for anticholinesterase, antioxidant, antiglucosidase and neuroprotective effect of three traditional medicinal plants for Alzheimer's Disease and Diabetes Mellitus dual therapy. *BMC Complement Altern Med.* 2018;18:77.
32. Pratico D. Oxidative stress hypothesis in Alzheimer's disease: a reappraisal. *Trends Pharmacol Sci.* 2008;29:609–615.
33. Firuzi O, Miri R, Tavakkoli M, Saso L. Antioxidant therapy: current status and future prospects. *Curr Med Chem.* 2011;18:3871–3888.
34. Murphy MP. Antioxidants as therapies: can we improve on nature? *Free Radic Biol Med.* 2014;66:20–23.
35. Fang L, Kraus B, Lehmann J, et al. Design and synthesis of tacrine–ferulic acid hybrids as multi-potent anti-Alzheimer drug candidates. *Bioorg Med Chem Lett.* 2008;18:2905–2909.
36. Sgarbossa A, Giacomazza D, di Carlo M. Ferulic acid: a hope for Alzheimer's disease therapy from plants. *Nutrients.* 2015;7:5764–5782.
37. Benchekroun M, Bartolini M, Egea J, et al. Novel tacrine grafted Ugi adducts as multipotent anti-Alzheimer drugs: a synthetic renewal in tacrine-ferulic acid hybrids. *Chem Med Chem.* 2015;10:523–539.
38. Chen Y, Sun J, Fang L, et al. Tacrine–ferulic acid–nitric oxide (NO) donor trihybrids as potent, multifunctional acetyl- and butyrylcholinesterase inhibitors. *J Med Chem.* 2012;55:4309–4321.
39. Tikhonova MA, Amstislavskaya TG, Belichenko VM, et al. Modulation of the expression of genes related to the system of amyloid-beta metabolism in the brain as a novel mechanism of ceftriaxone neuroprotective properties. *BMC Neurosci.* 2018;19:13.
40. Singh SK, Srivastav S, Yadav AK, Srikrishna S, Perry G. Overview of Alzheimer's disease and some therapeutic approaches targeting Abeta by using several synthetic and herbal compounds. *Oxid Med Cell Longev.* 2016;2016:7361613.
41. Nalivaeva NN, Belyaev ND, Turner AJ. New insights into epigenetic and pharmacological regulation of amyloid-degrading enzymes. *Neurochem Res.* 2016;41:620–630.
42. Wang DS, Dickson DW, Malter JS. beta-Amyloid degradation and Alzheimer's disease. *J Biomed Biotechnol.* 2006;2006:58406.
43. Grimm MO, Mett J, Stahlmann CP, et al. Neprilysin and Abeta clearance: impact of the APP intracellular domain in NEP regulation and implications in Alzheimer's disease. *Front Aging Neurosci.* 2013;5:98.
44. Van Kampen JM, Kay DG. Progranulin gene delivery reduces plaque burden and synaptic atrophy in a mouse model of Alzheimer's disease. *PLoS One.* 2017;12:e0182896.

45. Shirotani K, Tsubuki S, Iwata N. Neprilysin degrades both amyloid β peptides 1±40 and 1±42 most rapidly and efficiently among thiorphan- and phosphoramidon-sensitive endopeptidases. *J Biol Chem.* 2001;276:21895–21901.
46. Iwata N, Tsubuki S, Takaki Y, et al. Metabolic regulation of brain Aβ by neprilysin. *Science.* 2001;292:1550–1552.
47. Spencer B, Marr RA, Rockenstein E, et al. Long-term neprilysin gene transfer is associated with reduced levels of intracellular Abeta and behavioral improvement in APP transgenic mice. *BMC Neurosci.* 2008;9:109.
48. Kudrna JJ, Ugen KE. Gene-based vaccines and immunotherapeutic strategies against neurodegenerative diseases: potential utility and limitations. *Hum Vaccines Immunother.* 2015;11(8):1921–1926.
49. Merkel SF, Andrews AM, Lutton EM, et al. Trafficking of AAV vectors across a model of the blood-brain barrier; a comparative study of transcytosis and transduction using primary human brain endothelial cells. *J Neurochem.* 2017;140:216–230.
50. Krol S, Macrez R, Docagne F, et al. Therapeutic benefits from nanoparticles: the potential significance of nanoscience in diseases with compromise to the blood brain barrier. *Chem Rev.* 2013;113:1877–1903.
51. Hajipour MJ, Santoso MR, Rezaee F, et al. Advances in Alzheimer's diagnosis and therapy: the implications of nanotechnology. *Trends Biotechnol.* 2017;35:937–953.
52. Bungart BL, Dong L, Sobek D, et al. Nanoparticle-emitted light attenuates amyloid-β-induced superoxide and inflammation in astrocytes. *Nanomedicine.* 2014;10:15–17.
53. Sastre M, Ritchie CW, Hajji N. Metal ions in Alzheimer's disease brain. *JSM Alzheimer's Dis Rel Dement.* 2015;2:1014–1019.
54. Li M, Guan Y, Ding C, et al. An ultrathin graphitic carbon nitride nanosheet: a novel inhibitor of metal induced amyloid aggregation associated with Alzheimer's disease. *J Mat Chem B.* 2016;4:4072–4075.
55. Geng J, Li M, Wu L, et al. Mesoporous silica nanoparticle-based H2O2 responsive controlled-release system used for Alzheimer's disease treatment. *Adv Healthc Mater.* 2012;1:332–336.
56. Betzer O, Shilo M, Opochinsky R. The effect of nanoparticle size on the ability to cross the blood-brain barrier: an *in vivo* study. *Nanomedicine (Lond).* 2017;12:1533–1546.
57. Shilo M, Motiei M, Hana P. Transport of nanoparticles through the blood-brain barrier for imaging and therapeutic applications. *Nanoscale.* 2014;6:2146–2152.
58. Meir R, Popovtzer R. Cell tracking using gold nanoparticles and computed tomography imaging. *Wiley Interdiscip Rev Nanomed Nanobiotechnol.* 2018;10.
59. Betzer O, Meir R, Dreifuss T. *In-vitro* optimization of nanoparticle-cell labeling protocols for *in-vivo* cell tracking applications. *Sci Rep.* 2015;5:15400.
60. Agrawal M, Ajazuddin, Tripathi DK, et al. Recent advancements in liposomes targeting strategies to cross blood-brain barrier (BBB) for the treatment of Alzheimer's disease. *J Control Release.* 2017;260:61–77.
61. Rip J, Chen L, Hartman R. Glutathione PEGylated liposomes: pharmacokinetics and delivery of cargo across the blood-brain barrier in rats. *J Drug Target.* 2014;22:460–467.
62. Do TD, Ul Amin F, Noh Y. Guidance of magnetic nanocontainers for treating Alzheimer's disease using an electromagnetic, targeted drug-delivery actuator. *J Biomed Nanotechnol.* 2016;12:569–574.
63. Ansari SA, Satar R, Perveen A, Ashraf GM. Current opinion in Alzheimer's disease therapy by nanotechnology-based approaches. *Curr Opin Psychiatry.* 2017;30:128–135.
64. Amin FU, Hoshiar AK, Do TD, et al. Osmotin-loaded magnetic nanoparticles with electromagnetic guidance for the treatment of Alzheimer's disease. *Nanoscale.* 2017;9:10619–10632.
65. Kwon HJ, Kim D, Seo K, et al. Ceria nanoparticle systems for selective scavenging of mitochondrial, intracellular, and extracellular reactive oxygen species in Parkinson's disease. *Angew Chem Int Ed Engl.* 2018;57:9408–9412.
66. Dowding JM, Song W, Bossy K, et al. Cerium oxide nanoparticles protect against Aβ-induced mitochondrial fragmentation and neuronal cell death. *Cell Death Differ.* 2014;21:1622–1632.
67. Smith MA, Richey Harris PL, Sayre LM, Beckman JS, Perry G. Widespread peroxynitrite mediated damage in Alzheimer's disease. *J Neurosci.* 1997;17:2653–2657.
68. Knott AB, Bossy-Wetzel E. Nitric oxide in health and disease of the nervous system. *Antioxidants Redox Signal.* 2009;11:541–554.
69. Guix FX, Ill-Raga G, Bravo R, et al. Amyloid-dependent triosephosphate isomerase nitrotyrosination induces glycation and tau fibrillation. *Brain.* 2009;132:1335–1345.
70. Lustbader JW, Cirilli M, Lin C, et al. ABAD directly links aß to mitochondrial toxicity in Alzheimer's disease. *Science.* 2004;304:448–452.
71. Schmidt C, Lepsveridze E, Chi S, et al. Amyloid precursor protein and amyloid-β-peptide bind to ATP synthase and regulate its activity at the surface of neural cells. *Mol Psychiatr.* 2008;13:953–969.
72. Du H, Guo L, Fang F, et al. Cyclophilin D deficiency attenuates mitochondrial and neuronal perturbation and ameliorates learning and memory in Alzheimer's disease. *Nat Med.* 2008;14:1097–1105.
73. Shvedova AA, Castranova V, Kisin ER, et al. Exposure to carbon nanotube material: assessment of nanotube cytotoxicity using human keratinocyte cells. *J Toxicol Environ Health.* 2003;66:1909–1926.
74. Shvedova AA, Kisin ER, Mercer R, et al. Unusual inflammatory and fibrogenic pulmonary responses to single-walled carbon nanotubes in mice. *Am J Physiol Lung Cell Mol Physiol.* 2005;289:L698–L708.
75. Lam CW, James JT, McCluskey R, Hunter RL. Pulmonary toxicity of single-wall carbon nanotubes in mice 7 and 90 days after intratracheal instillation. *Toxicol Sci.* 2004;77:126–134.
76. Cui D, Tian F, Ozkan CS, Wang M, Gao H. Effect of single wall carbon nanotubes on human HEK293 cells. *Toxicol Lett.* 2005;155:73–85.

77. Muller J, Huaux F, Moreau N, et al. Respiratory toxicity of multi-wall carbon nanotubes. *Toxicol Appl Pharmacol.* 2005;207:221–231.
78. Heller D, Baik S, Eurell T, Strano M. Single-walled carbon nanotube spectroscopy in live cells: towards long-term labels and optical sensors. *Adv Mater (Weinheim, Ger).* 2005;17:2793–2799.
79. Sato Y, Yokoyama A, Shibata K, et al. Influence of length on cytotoxicity of multi-walled carbon nanotubes against human acute monocytic leukemia cell line THP-1 *in vitro* and subcutaneous tissue of rats *in vivo. Mol Biosyst.* 2005;1:176–182.
80. Bottini M, Bruckner S, Nika K, et al. Multi-walled carbon nanotubes induce T lymphocyte apoptosis. *Toxicol Lett.* 2006;160:121–126.
81. Smart SK, Cassady AI, Lu GQ, Martin DJ. The biocompatibility of carbon nanotubes. *Carbon.* 2006;44:1034–1047.
82. Kagan VE, Tyurina YY, Tyurin VA, et al. Direct and indirect effects of single walled carbon nanotubes on RAW 264.7 macrophages: role of iron. *Toxicol Lett.* 2006;165:88–100.
83. Amatore C, Arbault S, Cristhina D, et al. Stimulates or attenuates reactive oxygen and nitrogen species (ROS, RNS) production depending on cell state: quantitative amperometric measurements of oxidative bursts at PLB-985 and RAW 264.7 cells at the single cell level. *J Electroanal Chem.* 2008;615:34–44.
84. Garza KM, Soto KF, Murr LE. Cytotoxicity and reactive oxygen species generation from aggregated carbon and carbonaceous nanoparticulate materials. *Int J Nanomed.* 2008;3:83–94.
85. Lacerda L, Soundararajan A, Pastorin G, et al. Dynamic imaging of functionalized multi-walled carbon nanotube systemic circulation and urinary excretion. *Adv Mater (Weinheim, Ger).* 2008;20:225–230.
86. Yang Z, Zhang Y, Yang Y, et al. Pharmacological and toxicological target organelles and safe use of single-walled carbon nanotubes as drug carriers in treating Alzheimer disease. *Nanomedicine.* 2010;6:427–441.
87. Kunzmann A, Andersson B, Thurnherr T, Krug H, Scheynius A, Fadeel B. Toxicology of engineered nanomaterials: focus on biocompatibility, biodistribution and biodegradation. *Biochim Biophys Acta.* 2011;1810:361–373.
88. Giordano C, Albani D, Gloria A, et al. Nanocomposites for neurodegenerative diseases: hydrogel-nanoparticle combinations for a challenging drug delivery. *Int J Artif Organs.* 2011;34:1115–1127.
89. Billi F, Campbell P. Nanotoxicology of metal wear particles in total joint arthroplasty: a review of current concepts. *J Appl Biomater Biomech.* 2010;8:1–6.
90. Saraiva C, Praça C, Ferreira R, Santos T, Ferreira L, Bernardino L. Nanoparticle-mediated brain drug delivery: overcoming blood-brain barrier to treat neurodegenerative diseases. *J Contr Release.* 2016;235:34–47.
91. Mahmoudi M, Quinlan-Pluck F, Monopoli MP, et al. Influence of the physiochemical properties of superparamagnetic iron oxide nanoparticles on amyloid β protein fibrillation in solution. *ACS Chem Neurosci.* 2013;4:475–485.
92. Saraiva AM, Cardoso I, Pereira MC, et al. Controlling amyloid-β peptide(1–42) oligomerization and toxicity by fluorinated nanoparticles. *Chembiochem.* 2010;11:1905–1913.
93. Cabaleiro-Lago C, Quinlan-Pluck F, Lynch I, Dawson KA, Linse S. Dual effect of amino modified polystyrene nanoparticles on amyloid β protein fibrillation. *ACS Chem Neurosci.* 2010;1:279–287.
94. Ma Q, Wei G, Yang X. Influence of Au nanoparticles on the aggregation of amyloid-β-(25–35) peptides. *Nanoscale.* 2013;5:10397–10403.
95. Ghavami M, Rezaei M, Ejtehadi R, et al. Physiological temperature has a crucial role in amyloid beta in the absence and presence of hydrophobic and hydrophilic nanoparticles. *ACS Chem Neurosci.* 2013;4:375–378.
96. Wu WH, Sun X, Yu YP, et al. TiO_2 nanoparticles promote β-amyloid fibrillation *in vitro. Biochem Biophys Res Commun.* 2008;373:315–318.
97. Moore KA, Pate KM, Soto-Ortega DD, et al. Influence of gold nanoparticle surface chemistry and diameter upon Alzheimer's disease amyloid-β protein aggregation. *J Biol Eng.* 2017;11:5.
98. Puschmann TB, de Pablo Y, Zanden C, Liu J, Pekny M. A novel method for three-dimensional culture of central nervous system neurons. *Tissue Eng C Methods.* 2014;20:485–492.
99. Yin T, Yang L, Liu Y, Zhou X, Sun J, Liu J. Sialic acid (SA)-modified selenium nanoparticles coated with a high blood-brain barrier permeability peptide-B6 peptide for potential use in Alzheimer's disease. *Acta Biomater.* 2015;25:172–183.
100. Wong HL, Wu XY, Bendayan R. Nanotechnological advances for the delivery of CNS therapeutics. *Adv Drug Deliv Rev.* 2012;64:686–700.

Index

Note: Page numbers followed by "f" indicate figures, "t" indicate tables.

Printed in the United States
By Bookmasters